imagine *this*

RIT's Innovation + Creativity Festival

imagine *this*

RIT's Innovation + Creativity Festival

Edited by BRUCE A. AUSTIN

Photography edited by A. SUE WEISLER

imagine *this* RIT's Innovation + Creativity Festival

Copyright © 2018 Rochester Institute of Technology

All rights reserved. No part of this book may be reproduced in any form or by any mechanical or electronic means without written permission of the publisher and/or individual contributors, except in the case of brief quotations embodied in critical articles and reviews.

Published and distributed by:
RIT Press
90 Lomb Memorial Drive
Rochester, New York 14623-5604
http://ritpress.rit.edu

Printed in the U.S.
ISBN 9781939125484

Library of Congress Cataloging-in-Publication Data

Names: Austin, Bruce A., 1952- editor. | Rochester Institute of Technology issuing body.
Title: Imagine this : RIT's innovation + creativity festival / edited by Bruce A. Austin.
Description: Rochester, New York : RIT Press, [2018]
Identifiers: LCCN 2017055888 | ISBN 9781939125484 (print (hardcover) : alk. paper)
Subjects: LCSH: RIT's Imagine Festival--History. | Technology--New York (State)--Rochester--Exhibitions--History.
Classification: LCC T397.R63 I43 2018 | DDC 607/.3474789--dc23
LC record available at https://lccn.loc.gov/2017055888

CONTENTS

ix   **Foreword**
BRUCE A. AUSTIN

1   1   **Inventing**
BARRY CULHANE

2   15   **More than an Event**
ANDREW QUAGLIATA

3   55   **Promoting the Experience**
BOB FINNERTY

4   83   **Maturation and Momentum**
HEATHER COTTONE

5   117   **At the Intersection of Left and Right Brain**
JEFF SPEVAK

6   145   **A Galvanizing Force**
BILL DESTLER

FOREWORD

"**Chutzpah!**" A word not requiring punctuation, and never a contender for the name of the ambitious if ambiguously defined one-day spring festival offered at RIT beginning in 2008. I have this on good, solid authority. Though possessing, as it turned out, the virtue of accuracy, its connotation is a bit cheeky for an event associated with a university. But the word captures the audacity and aspiration of what planners envisioned as the scope of Rochester Institute of Technology's annual festival of creativity and innovation.

"Imagine" is evocative and resonates, even if it is not the first instance of the word's use as a proper noun. Most famously, John Lennon's song (1971). "Imagine" is lyrical without insisting upon familiarity with the lyrics, democratic insofar as it permits interpretative breadth, and universally inspirational. Effortlessly memorable, "Imagine" paradoxically commands as much as it defies obedience. A celebratory extension of RIT's historic amalgamation between the Mechanics Institute, where things were invented and made, and the Rochester Athenaeum, where ideas were discovered and discussed. Imagine RIT is a 21st-century manifestation of its 19th-century heritage.

Imagine RIT has the spirit, albeit not the scale, of 19th-century World's Fairs. Then, as now, such festivals were billboards for the marvels of the age, forecasting artistic and technological developments. Beginning with London's Crystal Palace in 1851, followed by the Philadelphia Centennial (1876) and the Columbian World's Fair (1893, Chicago) and years forward, fairs introduced new materials such as

FOREWORD

aluminum; media, including the telephone; processes, such as electroplating silver; products ranging from Cracker Jack to fluorescent lightbulbs; and creativity by way of art pottery. Attendance ensured a firsthand, unmediated experience. RIT's own connection to a World's Fair is the seven Harry Bertoia "dandelion" sculptures located at various places on campus. They were originally displayed at the 1964-65 New York World's Fair in Flushing Meadows Park, Queens. *Life* magazine's (May 1, 1964) story on the Fair presented a photograph of them being sprayed with water and situated in a round reflecting pool at the Kodak pavilion.

Imagine RIT is rooted in the community and Rochester's summertime festival bookends: The Corn Hill and Clothesline festivals. Populated first by Bohemian beatniks and later by hippies, each offers work from one end of the continuum to the other: craft to art with liberal mixing of the two. And the ties to both extend to RIT's curricular interests: the founding of the Department of Fine and Applied Arts

At the School of Animation studio inside Gannett Hall, guests walking onto the giant "green screen" are transported to the visual magic of Hollywood and superimposed into many exciting locations, including floating in outer space or swimming in a stream.
Photo by Brett Carlsen, 2014

by Theodore Hanford Pond in 1902, and the arrival of the School for American Craftsmen (SAC) in 1950, thanks to Aileen Osborn Vanderbilt Webb. The Bauhaus-inspired architectural style of RIT's 1968 Henrietta campus, literally edgy, is mirrored by the pedagogical practice inside SAC classrooms: Students studied with a craft master and an art master. Ideas and tools, working in concert. Inside and outside, RIT is synonymous with innovation.

Launching the first Festival presented organizers with challenges: Explaining the mysterious event to potential exhibitors, publicizing the unknown entity, and persuading people to come to campus in late spring when the weather is "iffy" and the Festival's offerings amorphous. Andrew Quagliata, organizer for the first four Festivals, and Bob Finnerty, who marketed all 10, explain the problem-solving process. Within a few years, and several fewer than expected, Imagine RIT's success was established. Sustaining and enhancing the Festival were the problems met by

Electric sparks from a Van de Graaff generator, Simone Center for Innovation and Entrepreneurship. Photo by Matt Wittmeyer, 2010

Heather Cottone who, as this book goes to press, is organizing her seventh Imagine RIT. Acknowledging how Rochester's outdoor festivals "elbow each other for space on the calendar," Jeff Spevak contextualizes Imagine RIT in Rochester's jam-packed festival season while using the campus's tallest sculpture, Albert Paley's *Sentinel*, as emblematic of the RIT ethos.

Photographs of Imagine RIT are extensive and comprehensive, as would be expected at a place with a school of photography. Because the Festival occupies such a large physical space on campus, three "zones"—each a slice of the campus—were established, within which individual photographers worked. Up to three freelance photographers, often RIT alumni, along with three RIT staff photographers, captured the action at each Festival. RIT staff photographers were Mark Benjamin, Elizabeth Torgerson-Lamark, and A. Sue Weisler. Having photographed all 10 Imagine RIT Festivals, Sue Weisler serves the present volume as photography editor. From the more than 5,000 "official" images created over the past decade, Sue curated the selection shown between these covers.

A palpable energy and infectious enthusiasm exist at Imagine RIT. Exhibitors have it—and it is not the "show-off" hubris variety, but instead the spirit of sharing something really neat and wanting to make sure everyone gets a chance to experience it. Volunteers embrace it; they know they are part of something by now well-defined, yet still filled with delightful surprise and discovery. And those who attend look forward to it, because it remains fresh and new and novel every single year. The competition at Imagine RIT, if indeed it exists, is about who is going to have the most fun, learn the most, and become most engaged. A sport at which there genuinely are no losers. Imagine RIT's spunk and spirit make it difficult to be anything but a fan and a booster. There is a willingness and inspiration to try, coupled with the conviction of possibility.

A tweet before there was Twitter, a meme before that expression became a cliché, Imagine RIT is an assertion, an announcement, a nonverbal verb. Optimistic and aspirational, like a World's Fair. Fun, scrappy, and maybe even a little frisky, promising a peek at real things usually reserved for fiction. Imagine RIT, a three-dimensional festival without boundaries, and after 10 years, a part of the warp and weft woven into and extending the Institute's fabric into the 21st century. Imagine. Indeed.

Acknowledgments: Lynn Wild supported this project enthusiastically as she has many others. Colleagues Mike Johansson and Rudy Pugliese each read and commented on texts; Marnie Soom advised on design. Barry Culhane shepherded the proposed book's budget to President Destler's office for approval. Andrew Quagliata, Bob Finnerty, Heather Cottone, and Jeff Spevak eagerly agreed to their "assignments" and were responsive to editing and questions. Kathleen S. Smith was a wordsmith partner and the book's diligent copy editor. Sue Weisler and her trained, sensitive eye plowed through a nearly unimaginable number of photographs, providing Lisa Mauro with multiple image choices, which she thoughtfully integrated into her appealing design for the book. Melinda Beyerlein was my reliable support assistant on this project, as she was for countless projects over more than a decade. From picky but essential details to broader, and sometimes tedious, assignments, she cheerfully and professionally gave her deeply invested attention.

Bruce A. Austin
Director, RIT Press

"Making the Invisible Visible," College of Imaging Arts and Sciences. Photo by A. Sue Weisler, 2012

FOREWORD

Mark Benjamin, 2012

A. Sue Weisler, 2012

imagine *this*   xvii

SLIM — See-through, Life-size Interactive Monitor — allows information to be shown on both sides of the screen boards and face-to-face communication is maintained as writing, showing illustrations and videos, and pointing to images are performed on the board.
Photo by Mark Benjamin, 2013

Mark Benjamin, 2012 and 2014

A. Sue Weisler, 2009

1 INVENTING

BARRY CULHANE

Bill Destler had been president of RIT for just two months when he called me into his seventh-floor office and threw out an idea as expansive as the campus sprawling below us. What did I think about holding an annual community event—we called it several different names before settling on "festival"—to showcase the "left" and "right" brain talents of our students, faculty, and staff members?

We tossed out ideas and I scribbled notes as we began to *invent* what such a festival might look like. Community focused. Open to the public. Family friendly. Free admission and parking. Heavy emphasis on technological innovations as well as those in visual and performing arts. In other words, the best RIT has to offer.

The idea itself was affirming and more than a bit daunting, particularly when, for some reason, we decided to hold the event in nine short months. We struggled for a festival moniker until one morning on the car radio, John Lennon's "Imagine" greeted me on my way into work. I raced into Bill's office and said, "I've got it: 'Imagine.'" He immediately shot back, "Imagine *RIT*" and a tradition was born.

We built an organizing group of administrators to reach out to academic and staff department heads across campus, lined up a corporate sponsor, laid the groundwork for an anticipated crowd of perhaps several thousand visitors…and stood back and got out of the way, providing logistics and marketing as needed. We quickly realized that when people are invested in an idea, they find a way to make things work. And the idea for Imagine RIT clearly had hit a "Tiger Pride" nerve.

INVENTING

Of course, it wasn't all smooth sailing. There was some confusion and minor territorial resistance on campus. Communication occasionally broke down and deadlines weren't always met. Our unofficial motto became: "If things go wrong, don't blame. Recover. And don't do it that way next time."

Probably one of our best moves that first year was to hand out "Tiger Orange" tote bags, the first of many successful giveaways that allowed Imagine RIT's visitors to leave campus with tangible evidence of RIT's intellectual and creative stature, and allowed us to promote RIT in a sustainable and visible way.

Our call for exhibit proposals generated every conceivable idea, from an automatic hot dog machine to nitrogen ice cream. We began a "Campus Countdown" to Festival Day to keep enthusiasm up. And we decided to have one outstanding exhibit from each of the colleges on display in the Gordon Field House, inadvertently leading to confusion among some visitors who thought that *all* the Imagine RIT exhibits

The Mobilized Robotic Hot Dog Assembler dispenses toppings to the customer's order, College of Applied Science and Technology.
Photo by A. Sue Weisler, 2008

were housed there. In March of that inaugural year, without a shred of evidence, I boldly predicted Festival Day weather—sunny with no chance of rain. And, luckily, I nailed it. Along with the warming sun, an estimated 17,000 people visited RIT May 3, 2008. Nine straight years of Festival-day clear skies followed. Students became convinced after the second year that President Destler had a secret weather machine installed on top of Albert Paley's sculpture, *Sentinel*, to guarantee good weather.

Today, Imagine RIT is the heart and soul of the University, with outcomes both expected and unexpected. No one anticipated, for example, how much internal pride the event generated among faculty, staff, and students. Or how wildly popular the event would be among pre-teens—until we heard over and over again from current students and alumni that they decided to attend RIT after visiting the Imagine RIT Festival. Turns out that hot dog making machine was worth its weight in recruiting gold.

Liquid nitrogen ice cream, College of Science.
Photo by Mike Bradley, 2016

INVENTING

More than 32,000 people attended Imagine RIT's 10th anniversary celebration in 2017. And for the first time, the skies darkened and it began to rain midway through the event. Visitors and exhibitors didn't abandon Imagine RIT when the weather turned, and we realized then the Festival's true "staying power."

People frequently forget the Imagine RIT name—"Imagination," "Image," and a dozen other iterations exist on social media. But the most important thing is that every year brings a new cohort of exhibitors presenting genuinely novel ideas, and every year, people mark their calendars and delightedly return to be amazed, educated, and inspired by what is Imagined at RIT.

Dr. Barry Culhane *is Executive Assistant to the President of Rochester Institute of Technology, has been Imagine RIT's Chairman since the Festival began, and has been an RIT faculty member since 1974.*

National Technical Institute for the Deaf gallery of exhibitions, learning to fingerspell one's name in American Sign Language.
Photo by Mark Benjamin, 2011

imagine *this* 5

Max Schulte, 2009

INVENTING

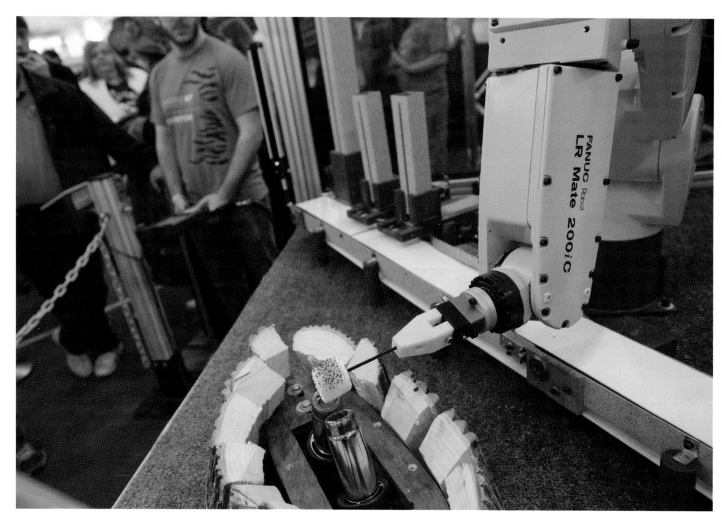

imagine *this*

The robotic S'Mores Experience machine cooks and assembles gooey, chocolatey s'mores for visitors. The robot was designed by students in the College of Applied Science and Technology and the treat presented for consumption in a custom-designed package.
Photos by A. Sue Weisler, 2014

8　INVENTING

Toyota Production Systems Lab, Kate Gleason College of Engineering, featuring reconfigurable production lines with storage areas and kitting areas, conveyors, and conveyance operations.
Photo by Brett Carlsen, 2014

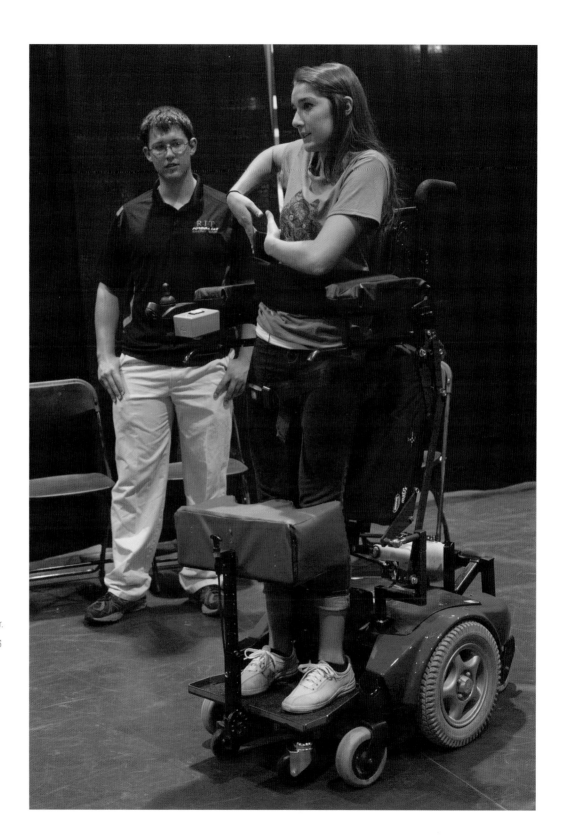

Motorized Stand-Up Electric Wheelchair.
Photo by A. Sue Weisler, 2016

imagine *this* 11

The prototype immersive classroom with wide-view, large projection display, wallpapers the learning environment using spectral information from the Van Gogh painting to simulate what it would look like under different light sources.
Photos by A. Sue Weisler, 2008

Brett Carlsen, 2014

A. Sue Weisler, 2009

Mike Bradley, 2015

2 MORE THAN AN EVENT

ANDREW QUAGLIATA

Imagine RIT was intended as a presentation to others, an event showcasing RIT's thriving campus, populated by intellectually curious and creatively inclined students, faculty, and staff. The Festival's unanticipated bonus was internal: It enhanced cross-college collaborations, shaped curricular decisions, drew alumni back to campus, facilitated student recruitment, broadened RIT's brand, and initiated a tradition now extending beyond a decade. Most significantly, Imagine RIT empowered and inspired young people to see themselves as innovators and creators. The Festival energized and excited the next generation of scientists, mathematicians, engineers, and artists. But it didn't happen overnight.

Organizing large-scale events is not a novelty at RIT. Commencement perhaps is the most familiar example, as thousands of visitors, simultaneously, if briefly, descend on campus. But Imagine RIT was unlike Commencement, as in its inaugural year it was an event without a playbook. A 1,000-piece puzzle spread across a 1,300-acre card table. Puzzle pieces and participants likewise numbered in the thousands. And organizing for its debut occurred without an illustrated box to guide the puzzle's assembly. Moreover, instead of inanimate pieces, this puzzle's elements were vessels of kinetic energy and boundless potential.

The public face of Imagine RIT is its exhibits, interactive displays, and presenters. Visitors experience, observe, and engage with the creativity and innovation that Imagine RIT promises. Behind that face, a coordinated, strategically driven organization ensures the visitor's seamless, satisfying experience.

Rewind to August 2007. Just months after taking the helm as RIT's ninth president, Bill Destler presented an expansive vision for putting RIT's "unique program mix" on display. Barry Culhane, executive assistant to the president, embraced the leadership role as Imagine RIT's Chairman. President Destler neither micromanaged the event nor limited what it would become. Imagine RIT began with a blank canvas. Quickly, event organizers learned that creating something new and on the scale envisioned was equal parts privilege and challenge.

Despite the short lead time, RIT was primed to launch such an event. With more than 175 years of history and half a century after awarding its first baccalaureate degrees, all the necessary elements for Imagine RIT were in place, poised to act in support of the bold venture: faculty energetically engaged in teaching and productive in scholarship, enthusiastic students eager to share their talents, and a "can-do" staff with "will-do" abilities.

Max Schulte, 2010

Culhane sought recommendations for key people to serve on an overarching Imagine RIT Planning Group. Members included Kim Slusser, Bob Finnerty, Ryan Giglia, Rebecca Johnson, Stan McKenzie, Kit Mayberry, Melody Cofield, Cindee Gray, Ed Lincoln, David Pankow, Reese Le Guerrier, Rod Lezette, Gary Gasper, Bonnie Meath-Lang, Lee Twyman, Terry Walker, and Heather Cottone. This group played a role in shaping every element of what Imagine RIT was to become. Their foundational decisions ranged from choosing the event date and name to the logo design and the decision not to sell merchandise. They anticipated obstacles and became Festival cheerleaders. The Planning Group and four task-specific committees were the edge pieces for the Imagine RIT puzzle. They spent hundreds of hours sketching the inaugural event, long before I was hired as Manager of Special Projects for Imagine RIT in February 2007, a role I held through the fourth Festival in 2012.

Tom Brenner, 2014

Packing Some STEAM

The 2015 Imagine RIT Festival featured more than 400 interactive exhibits in Science, Technology, Engineering, the Arts, and Mathematics (STEAM).

Visitors got a preview of the XYZ Camera Rig, which was preparing to move to its new home at the Metropolitan Museum of Art in New York City later in the month. "We took over where last year's design team left off with the first model of the rig," said Fabrice Bazile, a fifth-year Electrical Engineering student.

Bazile was one of six students on the project team that re-designed a mobile camera rig system for the museum to image larger works of art that cannot be moved easily. The rig can move horizontally across the floor, and its central system can be raised vertically to nearly 18 feet. The design team also improved the vertical lift to provide more stability for the mobile camera rig and system that the museum staff used to image rooms of artwork such as tapestries and other pieces from the museum's permanent collection. In the festival's Computer Zone, a "Pulse" exhibit allowed audiences to listen to a short concert of classical and pop music while their heartbeats were visualized on a large projector screen.

After downloading an Android or iOS app created by Computer Science students, users slipped on a specially designed pulse sensor, created by Industrial Design students, that communicated with a mobile device to show heart rate, in addition to synchronized visuals created by New Media Design students.

The camera sensor was converted to be sensitive to infrared, allowing RIT photographers to take pictures without a filter to make the images. The resulting image reflected infrared light in ways that made dark glasses clear and other colors look rather surreal.

It took 50,000 K'nex pieces to build the Theme Park Engineering exhibit and seven large tables to display it in the Gordon Field House. The exhibit, which featured 20 twirling, swirling replicas of actual amusement park rides—including two miniature roller coasters and a Ferris wheel—was the creation of RIT's Theme Park Enthusiasts Club. In addition to the lighted amusement rides—powered by motors, pulleys, and a touch of gravity—the exhibit had information explaining G-forces and listed possible careers in the theme park industry and RIT-offered majors that could lead to such jobs. The 10-person club assembled the structures over three weeks, with teams working in six-hour shifts.

Derrick Hunt, University News Services, May 2, 2015

Still, no matter how carefully, thoroughly, and exhaustively we thought things through, forecasting is imperfect and cannot accommodate the unexpected. Perhaps the most benign example was one "unauthorized exhibitor" at the first Festival: A young man hoisting a hand-lettered sign announcing, "Free Hugs." He paraded across campus, dispensing, as his sign indicated, hugs to all who were receptive.

More than 50,000 K'NEX pieces, spread across seven large tables, form the Theme Park Enthusiasts Club's exhibition in the Gordon Field House. Photo by A. Sue Weisler, 2015.

THE PUZZLE ASSEMBLERS

The 18-member Planning Group for Imagine RIT's inaugural year included the chairs of four Imagine RIT committees and liaisons to each of RIT's eight (now nine) colleges. Each committee was tasked with often intersecting, essential event functions whose components together composed the "machine" that would make Imagine RIT "run": Logistics, Fundraising, Volunteers, and Marketing. The organizational chart rapidly became a large and sometimes unruly three-dimensional spider web. Meetings required generously-sized spaces and ample time to work through details. As the years progressed, the size of this group shrank. For Imagine RIT's second year, it was half the size; by the fourth year, membership numbered five—the Festival Chair and four Committee Chairs.

Generating publicity for and promoting Imagine RIT was the *Marketing Committee's* charge. Especially for its debut, this was no easy task, since no one had ever experienced such a festival. Further, because there was neither video nor photographs from past events to illustrate what was in store for visitors, marketers had to rely on stock imagery to portray that which had not yet occurred. Bob Finnerty discusses the Marketing Committee's work in more detail in Chapter 3.

Bill Destler forecasted and we planned for thousands of Imagine RIT visitors, most strangers to our sprawling campus. A veritable army of hosts, greeters, and guides—mostly students, faculty, staff, retirees, and alumni—assisted visitors. Unlike Festival exhibitors, who displayed ideas and inventions, *Volunteers* extended hospitality and reliable information to visitors at Welcome Tents, Information Booths, and People Mover Stops. At the first Imagine RIT Festival, wearing a volunteer T-shirt, I stood on the Quarter Mile walkway between the Student Alumni Union and the dorms for 45 minutes answering questions: Where can I find the nearest restroom? Can you help me figure out where I parked? Is there a face-painting exhibit? Can anyone get a ride on the golf carts? I took a lot of notes.

imagine *this*

Julie Kang Evans, 2009

A surprisingly large number of children became separated from their parents; Festival volunteers helped reunite them. Before noon that day, Volunteer captain Sue Provenzano reported a child had lost his mother. We dispatched an RIT Public Safety officer. Minutes later, another volunteer radioed about a mother looking for her son. We thought the problem was solved—but it was a different mother and son who were separated. By the end of the day, Public Safety and volunteers had responded to more than a dozen lost child calls. At our post-Festival debrief, Public Safety recommended that we provide children with identification bracelets. For subsequent Festivals, we ordered 5,000 wristbands and issued them to parents arriving with children under the age of 12. Volunteers asked parents to write their cell phone number on the bracelet their child would wear. The bracelets did not prevent parents from losing their children; they did help us reconnect them.

A. Sue Weisler, 2008

imagine *this*

Julie Kang Evans, 2009

Though the program guides were plentiful for the first Imagine RIT, they could not sufficiently address all the way-finding needs of visitors, at least in part due to the expansiveness of RIT's campus. From a relatively modest-sized footprint as originally built in 1968, RIT had grown to include more than 200 buildings. We limited Imagine RIT's indoor exhibits to 25 buildings the first year, and sought to design each Festival so that visitors would have a great experience even if they never set foot inside a building.

To further facilitate the crowd's movements, and to give Imagine RIT a genuine festival "feel," we created visual connections between outdoor exhibits and those hosted inside buildings. Placement and balance between indoor and outdoor exhibits was achieved through visual cues such as Information Booths and food tents. In the Festival's third year, National Technical Institute for the Deaf (NTID) exhibitors moved to the more centrally located Student Alumni Union (SAU), rather than in their academic home on the east side of campus, to increase their visibility and reduce Imagine RIT's footprint by about a third of a mile—in one direction. That same year, the Center for Student Innovation opened on the west side of campus, providing showcase space for a wonderful variety of student entrepreneurial ideas. Nearby, in the fourth year, Global Village opened. In addition to stretching the physical boundaries of the Festival, such growth allowed exhibitors greater flexibility and helped ensure that no two Imagine RIT events were the same.

We knew that people would follow sounds; and since RIT is home to several musical groups, we set up two performance stages at the first Festival to coax visitors from one side of campus to the other. A west-side stage was in the Infinity Quad, bordered by the Colleges of Science, Engineering, Liberal Arts, and Imaging Arts and Sciences, as well as The Wallace Center. Another was located outside the SAU and adjacent to the Simone Plaza, site of Albert Paley's 70-foot *Sentinel* sculpture, a centrally located visual landmark. Many of RIT's a cappella student groups performed throughout the day. As well, the RIT Pep Band—40 students in RIT hockey jerseys—played on the steps of the Eastman Kodak Quad, midway between the two stages. Due to the positive comments we received about the live entertainment offerings, we increased the number of stages to four for Imagine RIT's second year.

Selected outdoor exhibits along main walkways subtly ushered visitors from place to place. For instance, one first-year outdoor exhibit was the Strongman Competition. RIT's Weightlifting Club proposed a series of events in which members would compete to move heavy items (e.g., huge truck tires) farther and faster than other members. Although initial response to the proposal was tepid, the Club persisted and eventually convinced organizers that they would follow proper safety

imagine *this* 25

The Strongman Competition, presented by the Weightlifting Club: RIT students pull old cars and lift heavy weights and tires; the exhibit also offers instruction on physics, including area of balance and center of gravity.
Photo by Matt Wittmeyer, 2010

protocols. The exhibit placed the weightlifters strategically on the access road between Clark Gymnasium and the Gordon Field House. Happily, large numbers of Imagine RIT visitors gathered for the event, lining the wall of the long bridge over the road and encouraging further exploration of the campus.

The Volunteer Committee's responsibilities for Imagine RIT's second year were folded into a new Program Committee, whose members worked from February to May soliciting exhibitor proposals, assigning exhibit locations, communicating expectations, approving the print program, preparing emergency plans, overseeing event assessment, and recruiting, assigning, and training volunteers.

An event the size of Imagine RIT is costly, and while everyone involved in the first event was eager to make it excellent, no one was allocated additional financial resources to make that happen. Some viewed Imagine RIT as an opportunity to

A. Sue Weisler, 2012

A. Sue Weisler, 2011

request significant funding; one exhibitor remarkably sought $20,000. Comprising staff from RIT's Development & Alumni Relations division, the *Fundraising Committee* solicited cash sponsorships and in-kind donations to support Imagine RIT; because of the generosity of such sponsors, cash gifts continue to this day to cover Festival costs.

Though most exhibitor funding requests were not granted, the Planning Group ensured that event-wide logistical expenses, such as table and chair rentals, would be paid for centrally. Exhibitor-specific expenses became decentralized and were paid through fundraising efforts by Development staff.

Event planners encountered multiple situations in which costs could have doubled their actual totals. One common exhibitor request was for large, flat-screen televisions on which to display work. A vendor's quote for TV rentals revealed that it would be less costly to purchase them. At a meeting with RIT's Finance Division, we were encouraged to accept the rental option, as there would not be sufficient resources to set up, store, and maintain 50 televisions in addition to other mounting Imagine RIT obligations. After the meeting, Barry Culhane and I walked by a lab filled with large, flat screen computer monitors. We immediately suggested that exhibitors use their desktop computer monitors to display their work, a decision that reduced costs by thousands of dollars.

Much of the Festival's behind-the-scenes work, invisible to most but necessary for all, was the responsibility of the *Logistics Committee*. By far the largest and most diverse committee, it included representatives from Access Services (American Sign Language Interpreting), Environmental Health & Safety, Facilities Management, Food Service, Information Technology Services (ITS), Parking & Transportation, Public Safety, Risk Management, and more. Conveniently, these groups were accustomed to working with one another.

Facilities Management coordinated rental of 450 tables, 1,000 chairs, and 30 tents. Parking & Transportation rented nine buses that delivered visitors to and from remote parking lots at Monroe Community College and Marketplace Mall. Sixty rented golf carts transported visitors around campus and 20 extra security officers directed thousands of vehicles to the 8,000 campus parking spaces and then saw them safely off campus at the end of the day. ITS secured 30 radios for event organizers and coordinated them on the same broadcast channel.

As the Logistics Committee realized the expanse of exhibitor internet needs, ITS staff grew concerned about how much internet activity would take place in the Field House; most exhibits required at least one connection. In 2008, wireless internet access was not available everywhere on campus and the extant Wi-Fi network could not handle the anticipated level of activity. Ultimately, the Field House was temporarily hard-wired, ensuring reliable internet access on the day of the event.

The Logistics Committee prepared a signage plan to anticipate how visitors would find their way around campus from the moment they arrived until the time they left. The plan was wonderfully detailed. However, I could not justify spending the $17,000 quoted by an external vendor to print the signs. Instead, we used a fraction of the committee's signage recommendations and staffed remaining locations with volunteers. Wearing bright orange shirts, the volunteers answered hundreds of visitor questions that signs would never have been able to answer, and handed out printed programs as well. The unintended but welcome consequence of this approach was an enhanced visitor experience.

The Logistics Committee met almost weekly leading up to the first Imagine RIT to coordinate efforts, support exhibitors, and accommodate guests. By Imagine RIT's second year, they met once a month; and by the third year, with the help of an intranet site, the group met face-to-face only three times. In all, 75 people across all five committees embraced the charge issued by their new president. But by early March, sixty days prior to the inaugural event, many questions remained. The work of many committees ground to a near-standstill until decisions about the Festival's program were made.

PIECING THE PUZZLE TOGETHER

Beginning in December 2007, Imagine RIT exhibit proposals were solicited from RIT faculty, staff, and students through email, flyers posted on campus bulletin boards, and personal contact. The call for proposals sought interactive presentations, hands-on demonstrations, exhibits, and research projects. Barry Culhane gave brief presentations to Student Government, Staff Council, and the Academic Senate, and he visited every College's faculty meeting at least once. He urged faculty to enlist their best students as part of exhibition teams to demonstrate, engage, and interact with visitors

imagine *this*

Eight Beat Measure, RIT's original all-male a cappella group.
Photos by A. Sue Weisler, 2013

interested in learning about their discoveries. The first Festival's appeal was, by necessity, a bit nebulous. In fact, a Festival tagline—"Can You Imagine RIT?"—was probably as enigmatic to organizers as it was to exhibitors and visitors.

But RIT was filled with latent exhibitors who already knew what was in store for Imagine RIT's guests. Their scientific discovery or innovative technology had already been made; or it would be by the time of the Festival. Their art was already created, even if it wasn't yet on canvas or some other medium. They needed only a vehicle through which to make their ideas known—Imagine RIT.

Campus response to the invitation for proposals was overwhelming. Submissions flooded the Festival Chair's inbox. The first step was to get all the information in one place. One of Barry Culhane's student assistants, Raymond Olivas, spent spring break helping me merge proposal data into one large file where we could

House-shaped maze promotes habitat awareness, presented by RIT's Habitat for Humanity. Photo by A. Sue Weisler, 2013

Kate Melton, 2016

Indiana Drones

An off-the-shelf drone customized by Rochester Institute of Technology students for archaeological surveys was on exhibit at the 10th annual Imagine RIT Festival in 2017.

The Multispectral Imaging Drone was located in The Think Tank zone in the Mobius Quad. Members of the senior design team from the Kate Gleason College of Engineering and the College of Science developed the drone as an archeological survey tool to locate potential artifacts prior to a dig. Data taken off the drone could be used to locate objects on the ground or buried a few meters below the surface. The idea began with project manager Leah Bartnik, a senior in RIT's Chester F. Carlson Center for Imaging Science, who grew up in West Seneca, N.Y., near the Penn-Dixie Fossil Park and Nature Reserve. While in high school, Bartnik worked at Penn-Dixie during the summers and fell in love with paleontology.

She wanted her senior capstone project to combine her interests in imaging science and paleontology, and to give her experience on an engineering team. Because RIT doesn't have a resident paleontologist, Bartnik used her senior capstone project in imaging science to customize an imaging system for archeologist William Middleton, associate professor in RIT's College of Liberal Arts. The team designed its drone to accommodate Middleton's search for Mayan ruins at an excavation site in Oaxaca, Mexico.

Bartnik designed the imaging system that clips on to the drone and measures chlorophyll, or the green pigment in vegetation. Her system combined a regular camera and near-infrared sensors to measure the green pigment reflected in different wavelengths of light. Low levels of chlorophyll are a clue that an object of interest might be obstructing the roots. The thermal camera in the imaging system provided additional information to support or disprove the hunch. "If there is a feature that emits heat at a different rate than the ground, and its position correlates with a feature that is stressing out the vegetation, there is probably something right at the surface or a few meters below," Bartnik said.

The imaging system is about the size of a tissue box and attaches to the custom-made gimbal built by mechanical engineers Ben McFadden and Gavin Bailey The carbon-fiber gimbal is the pivotal support mounted to the drone. The imaging system and the gimbal had to be integrated with the drone. The four electrical engineers on the team—Ryan Moore, Jeremy Sardella, Stephen Depot, and Arthur Svec—tied together the various software and control systems, including a flight planning system.

Imaging scientists, mechanical engineers, and electrical engineers approach problems from the perspective of their own disciplines, and the project's success depended on the students' ability to interrelate. The yearlong project gave them insight into what makes a successful team work.

Susan Gawlowicz, University News Services, April 24, 2017

review all the "puzzle pieces": Number of proposals submitted, funding requested, participation by college, and more. Then ITS created an online tool that made it easier to review submissions formally, and communicate systematically with exhibitors. As quickly as proposals were submitted, anxious proposers wanted to know whether their ideas had been approved and where on campus they would set up.

Initially, three criteria guided proposal selections. We wanted exhibits that would offer visitors a hands-on, interactive experience. Second, we prioritized interdisciplinary exhibits. Last, for coveted space in areas such as the Field House, we

imagine *this*

A. Sue Weisler, 2017

sought nearly equal representation among RIT's eight colleges. As we confirmed exhibits at the Field House, the Logistics Committee identified a fourth selection criterion: each must meet all safety requirements. One proposed Field House exhibit, for example, required a large tent to simulate darkness for a solar system project. While the idea was appealing initially, RIT Environmental Health and Safety informed us that tents were not permitted indoors, as they would render the Field House sprinkler system ineffective if a fire broke out. That exhibit was relocated and a different science exhibit took its place in the Field House.

As decisions were made about Imagine RIT's program, I shared detailed spreadsheets with the Logistics Committee. Relieved to see a program taking shape, the committee had concerns about how to make a visitor's experience positive on a campus with a full mile between east and west side exhibits. What would ordinarily be a 20-minute stroll could easily turn into an hour-long adventure when 15,000 other people were also roaming the campus. Inspired by the railroad that encircles Disney World's Magic Kingdom, we designed a "people mover route" that transported visitors quickly and comfortably across the inner perimeter of campus on golf carts. Even with this and many other options in place, post-Festival surveys revealed that some visitors felt the Festival venues were too spread out.

THE BEST-LAID PLANS
The evening before the first Festival, I wished pointlessly for a few more days to tie up loose ends. What will outdoor exhibitors do if it rains? Will our volunteers be able to answer visitors' questions? There was no shortage of second guessing. But after months of planning, the only thing left to do was lend a hand. From 7 until 10 a.m., I fielded questions by radio and cell phone, checking in with volunteer captains and point people in each college. Everyone was positive; no major problems were reported. In fact, the best part of the day for me was chatting up visitors as I drove them around campus in a golf cart.

There were, of course, a few problems that first year. Garbage cans overflowed, ATMs ran out of cash, and outside food locations sold out of many items early. Some problems proved easier to fix than others. Facilities Management devised a new waste management plan for Imagine RIT's second year and used the Festival as an opportunity to educate attendees about landfill waste and single stream recycling. The Controller's Office promised that ATMs would be fully stocked on Festival Day.

RITchie's Ice Cream Experience, College of Applied Science and Technology, teaches alternative sources of energy, rewarding visitors with a sundae.
Photo by A. Sue Weisler, 2011

But figuring out the proper number of food locations and menu mix remained a challenge all four years I was involved. People attending festivals want festival food, and we missed the boat that first year. RIT Catering did what it does well every other day of the year. But Imagine RIT was unlike anything they had ever experienced. Visitors unfamiliar with the campus did not know how to find the cafeterias, meaning long lines at the three outdoor food tents, which also ran out of bottled water early. In post-Festival surveys, visitors told us they wanted more food locations and greater menu variety; it was at this point that we began conversations with external vendors. Visitors' positive response to our partnership with Dinosaur Bar-B-Que the first year gave us a good idea of the direction in which we needed to go. We partnered with Mario's Italian Restaurant the second year and Abbott's Frozen Custard in the third year.

Matt Wittmeyer, 2010

imagine *this* 39

Matt Wittmeyer, 2010

After the first Festival, we realized how much parents saw the educational value of Imagine RIT for their children, so when we planned the second year, we created a special icon in the print program to denote "kid-friendly" exhibits. We also created an exhibit Scavenger Hunt, where children under 12 could earn an Imagine RIT patch if they visited at least one exhibit from each of five disciplines: science, technology, engineering, art, and math.

LISTENING TO THE EXPERTS

To understand Imagine RIT's impact, we gathered feedback the first year from almost 2,700 visitors, exhibitors, and volunteers. Overall, survey results were positive. Virtually all—97 percent—said they would attend Imagine RIT again. The most frequent complaint was that there was too much to see and not enough time to see it. Visitors wanted the event to be longer or even held over two days. We saw this as a good problem to have, though we never considered a two-day Festival, due to financial and human constraints. But we did extend Imagine RIT by one hour beginning the second year.

Some survey respondents expressed concern about congestion on campus. As organizers, we loved seeing large crowds; but some visitors were frustrated by the long lines. After attendance at the second Festival reached 25,000, I noted in an internal report, "After seeing the campus at 25,000, should we reconsider using attendance as a measure of success? Perhaps 30,000 would not make Imagine RIT a more successful event. Some may argue that increasing the attendance could adversely impact the quality of the experience for visitors." That theory was tested the following year, when we felt sure we were going to break the attendance record after excessive traffic caused the Festival website to crash the day before the event.

A successful event can breed even greater success, and perhaps make the Marketing Committee's job easier. However, as attendance increased, organizers spent more time resolving logistical challenges. With the help of talented student assistant Lauren Shapiro, the "Plan Your Day" tool on the Imagine RIT website attracted visitors interested in building their own customizable itineraries in advance. Before the page went down, it had logged an astonishing 40,000 views in one day. Imagine RIT attracted an estimated 32,000 visitors that year.

A. Sue Weisler, 2017

CONTINUING THE TRADITION

Imagine RIT's debut filled the RIT community with pride. A Class of 1976 alumnus emailed me the Monday after: "I was never prouder to be part of this incredible campus community." But we knew it would take a concerted effort to embed the event into the fabric of RIT. The week after Imagine RIT, some students in my Public Speaking class sheepishly admitted that they had stayed in their dorm rooms rather than take the short walk to see what Imagine RIT was all about—an honest reminder that a segment of RIT's community was not yet convinced that Imagine RIT was worth exploring. Too, we lose many exhibitors every year to graduation. Our survey of exhibitors from the first Festival asked, "Do you plan to submit a proposal to be an exhibitor next year?" Almost half responded favorably, but a third were undecided; the figures seemed low and high, respectively, to Festival planners. We determined

A. Sue Weisler, 2011

at that moment to be more intentional about creating a sustainable event. Now we educate students annually about Imagine RIT with video testimonials from student exhibitors telling their stories of the positive attention their exhibits received.

We learned that exhibitors wanted more information about what was expected of them. For the second Festival, we developed a communication plan that provided regular updates as the event drew closer. We offered detailed explanations to exhibitors regarding how to interact with and communicate complex information to the public, including young people. We made it easier to submit proposals by upgrading the proposal submission system. The two most tangible adjustments were giving each exhibitor an Imagine RIT Exhibitor T-shirt and providing lunch vouchers. Suddenly, participating in Imagine RIT became "the thing to do."

Students enrolled in RIT's Science and Technology Entry Program (STEP), a state-funded initiative providing academic enrichment and college and career exploration to 7th to 12th graders.
Photo by Mike Bradley, 2016

Long before the exhibits were taken down and the tents packed, we realized Imagine RIT was much more than an event. The internal RIT commitment to ensure its success is even more remarkable. Imagine RIT continues to be the University's flagship event because of the more than 2,000 exhibitors and 500 volunteers who showcase what makes RIT special. RIT's 2005-2015 Strategic Plan urged a stronger sense of community and interaction among RIT's campus constituencies. Though the Strategic Plan was not specific about how the goal should be accomplished, looking back, it becomes clear that Imagine RIT became the community-building tradition laid out in the Plan.

Andrew B. Quagliata *was Manager for Special Projects in RIT's Office of the President for the first four Imagine RIT Festivals. He is a Management Communication Lecturer at Cornell University's School of Hotel Administration. He wrote "University Festival Promotes STEM Education" for the* Journal of STEM Education: Innovations and Research, 2015, 16(3), 20.

A. Sue Weisler, 2015

imagine *this* 45

RIT Cycling Club, Human-Powered Bike Blender: pedal a bike, burn some calories, and make a healthy, delicious smoothie to cool yourself down.
Photo by Emily McKean, 2009

The Auburn Brothers Band.

Photo by A. Sue Weisler, 2009

Proof of Purchase perform at Global Village.
Photo by Mike Bradley, 2015

imagine *this*   49

Kate Melton, 2017

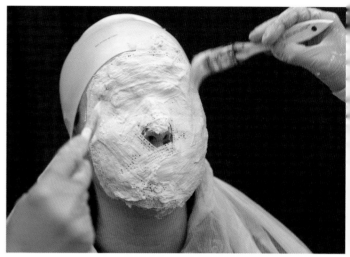

imagine *this*   51

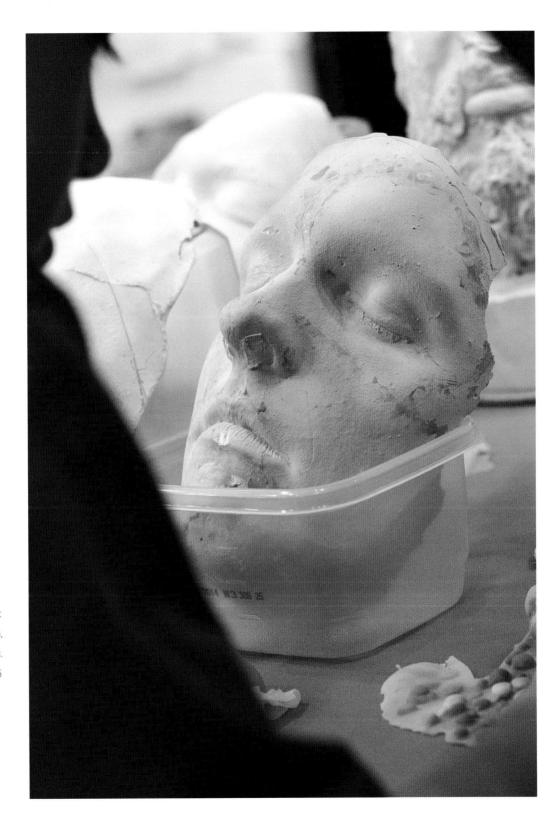

Prosthetics to Special Effects:
Life-Mask Creation,
College of Imaging Arts and Sciences.
Photos by A. Sue Weisler, 2014 and 2015

A. Sue Weisler, 2016

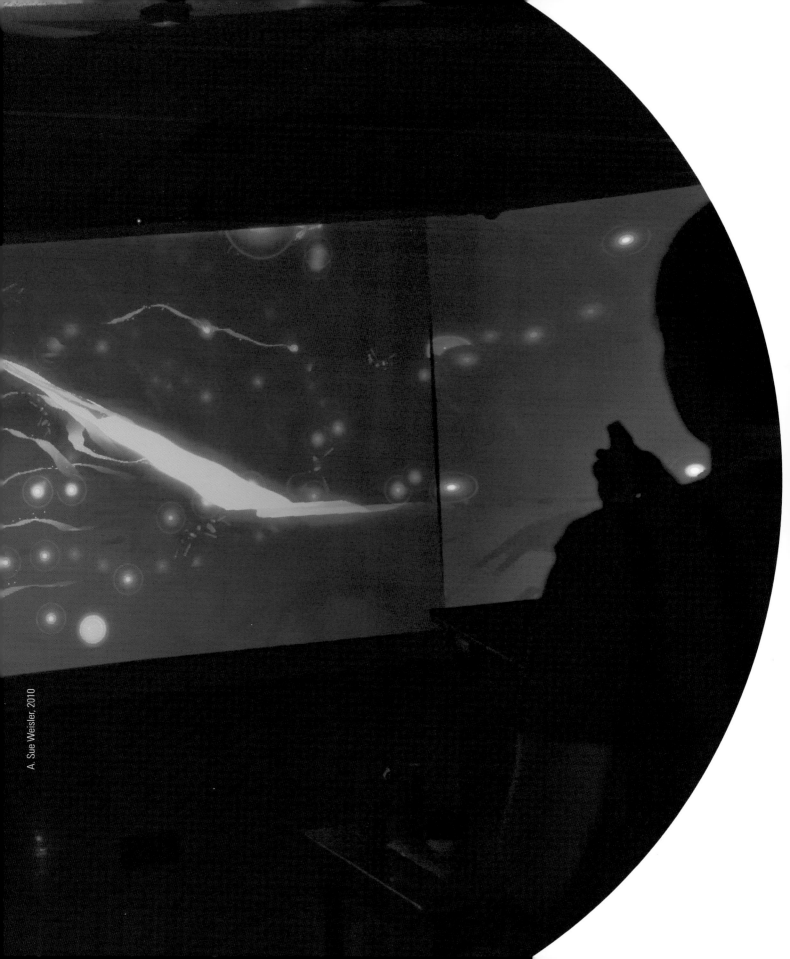
A. Sue Weisler, 2010

3 PROMOTING THE EXPERIENCE

BOB FINNERTY

Publicizing and promoting the inaugural launch of Imagine RIT presented a marketing challenge for the Festival's creators. Unlike most marketing initiatives, Imagine RIT in its infancy was neither product nor service—it was an "experience" and, in 2008, a novel one. How do you market and communicate creativity? Innovation? Should we take an abstract or practical approach? This was the opportunity presented to RIT's Marketing and Public Relations team as we began our efforts with both high expectations and a blank canvas.

On March 10, 2008, RIT's University News Services (UNS) was excited about a fresh story to tell. Inspired by a new president, the team was ready to roll out an event that would showcase all things "cool" about RIT, essentially a University Show and Tell on steroids. An afternoon news conference was planned at the Gordon Field House to unveil Imagine RIT: Innovation and Creativity Festival. For several months, the UNS team had worked tirelessly to prepare an illustrative sample of student and faculty Imagine RIT exhibitors for the conference. Projects included electric bicycles, a Formula One race car, hearing tests, a space tourism display, and an animatronic dog.

Any time you hold a news conference, it had better be for a good reason. With fewer resources and a shrinking press corps, journalists today are pressed for time. What do you have to tell the media that can't be accomplished in a phone call, an email, a news release, or even a Tweet? In the public relations business, your story must be good or you risk your professional credibility and your institution's reputation.

At the Field House, we were all set to give members of the media a taste of the upcoming Festival and entice them to want to attend. Exhibitors were ready. President Bill Destler's scripted remarks were ready. RIT's public relations team was ready. And four local TV stations, two newspapers, and one radio reporter pledged they were en route to learn about our new festival.

Until… a breaking news story derailed some of our efforts. New York Governor Eliot Spitzer, with his national reputation for convicting white-collar Wall Street

RIT Formula SAE race car.

Photo by A. Sue Weisler, 2016

criminals, was arrested for soliciting a prostitute in a Washington, D.C. hotel. Only two reporters—both clearly preoccupied with the emerging Spitzer story and in a hurry—showed up to our event. With the Spitzer story gaining momentum, we could sense they would not stay long, and that our story would be buried that night on the evening news.

"You can do everything right, have a great story, and make a great presentation. But at the end of the day, you're at the mercy of the news gods," recalled John Follaco, the UNS writer who worked on that first Imagine RIT media relations kit. The Imagine RIT Marketing Team would hold its collective breath for the next eight weeks as we waited for the first Festival to open May 3. We would have to work hard so that our tactics, besides just the initial news conference, would draw attention to the upcoming Festival.

Barry Culhane, Imagine RIT's Chairman, meets the press.
Photo by A. Sue Weisler, 2008

STAKE IN THE GROUND

In August 2007, newly appointed RIT President Bill Destler named Barry Culhane, executive assistant to the president, chairman of the Imagine RIT Festival. When Barry asked me to lead marketing and public relations for the event, I immediately accepted, recognizing the Festival's obvious potential to enhance the University's overall reputation. We had eight months to create, brand, and promote a festival from the ground up.

"Imagine RIT offers an opportunity to put a stake in the ground… to identify the University with innovation and creativity," Destler told the Imagine RIT Planning Group that month. He added: "The Festival's mission goes beyond showcasing the thriving RIT campus. Imagine RIT is designed to show what can be accomplished when smart, talented people with diverse strengths work together to solve complex societal problems. Innovation is one of our country's most competitive and strategic advantages."

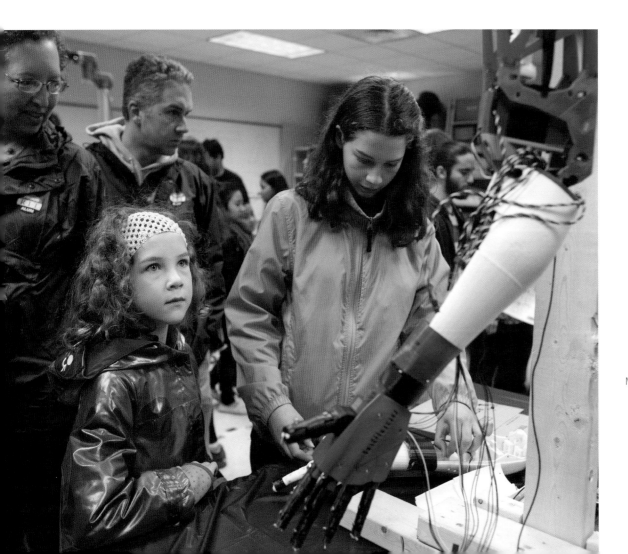

SPEX: Space Exploration Club
High Altitude Ballooning
College of Science
Photo by A. Sue Weisler, 2011

Mike Bradley, 2017

These inspiring words gave the Marketing Team a tangible messaging platform. President Destler repeatedly described RIT as a place where the "left brain and right brain collide"—the notion of melding analytical-methodical types such as engineers, scientists, and accountants with intuitive-creative types such as artists and dreamers. The President's goal was for the Festival to draw 30,000 attendees eventually, though he suspected it would take years for the event to become such a "top-tier" attraction.

Fall 2007 at RIT was spent planning the logistics of the event (as discussed in Andrew Quagliata's chapter) and developing a comprehensive marketing and communications plan. We needed a name, logo, and budget. I advocated for RIT Innovation and Creativity Jam, but the team and President landed on Imagine RIT after dozens of names were reviewed. Now that we had a name, we turned our attention to creating a logo, a mark that would be flexible enough to use on everything from posters to brochures to advertisements. Lead art director and designer Jeff Arbegast from University Publications came up with the logo, which positioned yellow, orange, and red balloon-like circles above a cloud. He wanted the logo to represent fun, but also to serve as an aspiration for Festival goers to explore and discover the best of what RIT had to offer.

Marketing had a budget of approximately $82,000 to develop and launch our campaign for the inaugural Imagine RIT. To stretch that budget, we worked with many partners, including Wahl Media (a media buyer for advertising, which keeps track of purchases and our mix of TV and radio ads), the Rochester *Democrat and Chronicle* daily newspaper, and the regional Time Warner Cable system. We honed our target audiences by focusing on families, including those of the 30,000 RIT alumni living in the Rochester region. Imagine RIT would engage them more deeply, we hoped, and get them back to their campus, now thriving with dozens of new buildings and expanded degree offerings. We also focused on regional middle-school students, believing that Imagine RIT would show them the exciting possibilities of pursuing degrees in STEM (science, technology, engineering, math) disciplines and the arts at RIT. And, of course, we wanted current students, faculty, and staff to get out of their residence halls and offices to see the full depth and breadth of "Brick City." Imagine RIT would demonstrate how engaging and exciting its disciplines are well beyond any textbook experience.

Mark Benjamin, 2014

Our first task as a Marketing Team was to identify the Festival's key messaging: "Discover what happens when innovation and creativity converge." "A festival showcasing the work of engineers and artists, entrepreneurs and designers, scientists and photographers." "Experience education and entertainment through a fusion of art and technology." In subsequent years, these messages and taglines became crisper and more direct, including "Imagine Your Possibilities," "What Will We Think of Next?" and "Experience the Future."

Spring 2008 was spent creating scripts for TV and radio commercials, developing advertisement copy for print publications and postcards, writing stories and news releases, producing videos, and identifying the right channels for placing and purchasing advertisements. The latter meant working with Wahl Media to identify which TV stations and shows would suit the right demographic for the Festival. We

A. Sue Weisler, 2009

imagine *this*   63

Max Schulte, 2009

placed 30-second television ads around shows such as *Wheel of Fortune, Jeopardy,* the *Today* show, *Oprah Winfrey,* and the hottest show at the time, *American Idol.* We diversified our advertising spending to reach minority communities by placing ads on WDKX, a popular local African-American radio station. And for mass appeal, we used a gimmick popular at the time—a "sticky note ad" placed on the front page of the *Democrat and Chronicle.* We also had web advertising on every major news site in town and placed hundreds of "table-tent cards" in the food courts of the three largest area shopping malls. All this advertising drove the public to a newly built website: www.rit.edu/imagine. Here, visitors could access descriptions of the 400 exhibits, get information on how to participate, see profiles of exhibitors, and watch numerous video features with students and faculty showing the concepts behind their exhibits.

Our commercials ran on four Rochester broadcast TV stations and the top five radio stations for 10 consecutive days leading up to the Festival. Print ads ran in the *Democrat and Chronicle, Rochester Business Journal, City Newspaper,* and RIT's weekly student-run *Reporter* magazine. So, despite the disappointment of the ill-timed news conference weeks earlier, we felt a genuine "buzz" that hinted that the Rochester community was aware of and enthusiastic about this new festival. And while it may seem elementary, we also emphasized four important words in all of our messaging—**"Free. Rain or Shine."** We were confident and ready. Let the inaugural Festival begin!

Early that morning, the campus looked a little sparsely populated; and most of the people I saw seemed to be RIT faculty and staff volunteers and their families. We were all anxious for crowds to materialize as we inched closer to the Festival's 10 a.m. start time. By 10:15 a.m., cars loaded with visitors were arriving; and by 10:30 a.m., we had a steady stream of cars, buses, and crowds pouring into campus. At noon, we saw traffic backed up at RIT's main entrance on Jefferson Road and knew then that Imagine RIT was on its way. "If you build it, they will come," I remember thinking. We *had* built it and they *were* coming—all 17,000 of them.

Kate Melton, 2017

LEARNING AND EXPERIMENTING WITH DIGITAL AND SOCIAL MEDIA

Having the words "innovation" and "creativity" in your product creates a bit more pressure to deliver that same ingenuity in your marketing, advertising, and public relations efforts. We knew that our work for the second Imagine RIT Festival had to be contemporary and thought-provoking. The good news: Planning became exponentially easier because we now had a marketing and PR template from which to work. More importantly, we now also had photos and videos to illustrate the amazing results from that first year. In Imagine RIT's inaugural year, we were hampered by having only stock archival images at our disposal for marketing efforts. Now we had real projects, real exhibits, real faces, and real testimonials from exhibitors and attendees to show off. But there was a new twist to contend with—the explosion of digital and social media. Facebook and Twitter had launched in 2006. Instagram followed in 2010 and Snapchat arrived in 2013.

As each new channel emerged, we adopted it, tweaking and experimenting along the way in our promotional campaigns. Digital advertising, for example, allowed us to highlight a science project for one audience and an arts-themed ad for another. Better yet, we now could blend the two.

We probably had the most fun using these new platforms in 2009, when Jared Lyon, a senior web developer in our Publications group, proposed creating a video for the upcoming Festival that he thought had a good chance to "go viral," which loosely meant garnering tens of thousands of views. Putting his domino collection to work on behalf of Imagine RIT, Jared meticulously set up 3,000 dominoes throughout the Marketing and Communications offices. In the video, a cardboard cut-out of President Destler "kicked" the first domino, causing a nearly two-minute chain reaction of interesting visual patterns that ended with the final domino tapping the "Print" button on a photocopier that spit out a large color version of the Imagine RIT logo. "Dominoes Everywhere" created the desired social media "buzz," with 250,000 views on Vimeo and nearly 94,000 on YouTube. Most gratifying for all of us, and particularly Jared, who spent hours setting up his display, Imagine RIT's attendance jumped by nearly 50 percent that year.

Movie Wizardry through Virtual Cinematography exhibit: a motion-capture system that translates the movements of an actual actor into a virtual movie studio within animated space.
Photo by A. Sue Weisler, 2017

> **Supertasters**
>
> Visitors to 2016's Imagine RIT Festival got the chance to find out if they were "supertasters," a feature in roughly 25 percent of the population who have more taste buds, have a stronger sense of taste, can taste some things others can't, and may be more picky eaters.
>
> It was part of an exhibit, "Exploring Psychology: How the Mind Plays Tricks," by psychology students from the College of Liberal Arts. Their tests illustrated how the brain can process information.
>
> They also had a test to trick the brain from remembering the names of colors, showed illusions, and had giveaways for children, including coloring books and brain-shaped erasers. "We're trying to get kids interested in the brain and psychology," said Assistant Professor Audrey Smerbeck of RIT's Psychology Department.
>
> For the supertaster test, visitors put small strips of paper on their tongues to see if they could taste PTC, short for phenylthiourea, which tastes very bitter to supertasters. The other 75 percent of the population tastes nothing. Women are more apt to be supertasters. "We evolved to taste bitter things to protect us from eating poisonous objects," Smerbeck said. "This helps us learn more about our sensory system, and how our tongue is used in different ways."
>
> In the other tests, visitors were asked to read a list of words that spelled various colors. The words themselves were in various colors that didn't match the words they spelled. The tester was told to read just the color of the list of words, not the word itself. "Your brain is distracted by what it focuses on," Smerbeck said. Preschoolers do well in this test because most haven't learned to read yet.
>
> Visitors also were asked to look at a picture. They may have seen Marilyn Monroe, or they may have seen Albert Einstein, depending on what they focused on and how far away they were from the image. "You focus on big things the farther you are from them, and smaller details the closer you are," she said.
>
> *Greg Livadas, University News Services, May 2, 2016*

In 2010, Imagine RIT's third year, the newly opened Simone Center for Student Innovation, at the west end of campus, gave the Marketing Team another great venue to promote and showcase exhibits. In the first two years, most of Imagine RIT's top exhibits were hosted inside the Gordon Field House, well to the east of the new building. The addition of the Simone Center now allowed us to create a "spine" of displays stretching across the campus, with a variety of new exhibits each year. When we reached President Destler's goal of 30,000 visitors by year three, he remarked, "I thought it would take 10 years to see crowds this size. We are ahead of schedule." Imagine RIT had truly arrived. Over the next seven years, crowds ranged in size from 25,000 to 35,000.

imagine *this* 69

For the Festival's first several years, the Marketing and PR Team targeted primarily areas that were a "day tripper's" drive from campus. These included Rochester, Buffalo, Syracuse, and Southern Tier markets such as Corning and Elmira. Our post-Imagine RIT surveys, however, revealed that we also were drawing visitors from Long Island, New York City, Cleveland, and Pittsburgh—often these were parents of students who were Imagine RIT exhibitors. Almost as often, an exhibitor's parents brought along a few siblings—whom we of course viewed as potential future RIT students.

The advent of paid social media advertising by 2010 not only allowed us to target prospective attendees by geographic region, but also by sex, age, location, and interests, such as art, music, technology, science, gaming, entertainment, and more. Social media advertising quickly became a valuable tool in our media mix toolbox, complementing our efforts in TV, radio, and print.

Matt Wittmeyer, 2010

In the Festival's fourth year, we began employing direct mail marketing and promotion. Ambitiously, we even sent an invitation to President Obama. He did not respond, unfortunately. But other elected officials, like New York State Senator Charles Schumer, paid visits. Direct mailings included newsletters, posters, and save-the-date magnets that were sent to middle schools, high schools, charter schools, and regional chapters of the Boy Scouts and Girl Scouts.

In 2011, we decreased our dependency on print advertising and increased our digital and social media "footprint" by adding mobile ads on smart phones, audio spots on Pandora, content-sponsored marketing website news stories, and video commercials on segmented smart TV channels and on YouTube, where in-house videos were reaching 2,000 to 5,000 views each year.

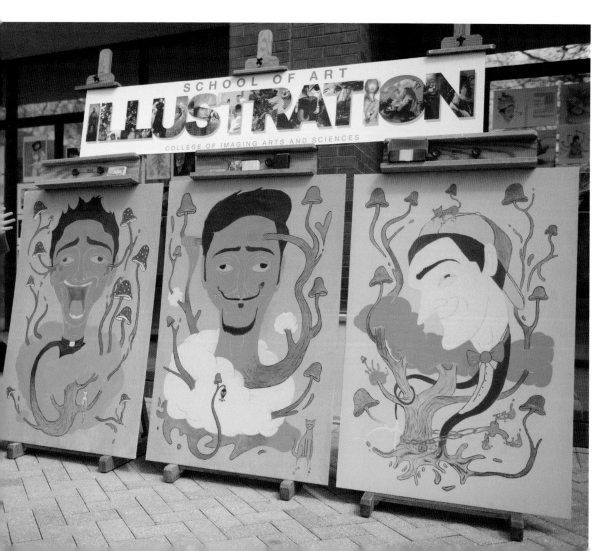

JuliAnna Patino, 2016

NIMBLE MARKETING

The first nine years of Imagine RIT, the Festival was amazingly rain free, despite Rochester's notoriously wet and chilly spring weather. Those first Festivals, to our delight, also were blessed with spectacular sunshine and 75-degree temperatures—perfect Festival weather. We knew, however, that the streak was bound to end one day, and just a week before the 10th anniversary Festival in 2017, it became clear that the Festival Day forecast was cloudy. Two days before the Festival, cold and rain hovered over Rochester and it seemed inevitable that our good fortune was about to end. Knowing the forecast a few days earlier, however, gave us just enough time to prepare and adjust our marketing efforts to counterbalance the predicted precipitation. What followed was an excellent example of how a flexible, creative team working with digital advertisers and social media outlets—along with a

Concrete canoe, College of Applied Science and Technology.
Photo by Sean Sullivan, 2014

fortuitous mistake by one advertising medium—helped us leverage our message to remind would-be Festival goers that the show would go on, "Rain or Shine."

First, we changed our digital advertising and social media messaging to pump up Imagine RIT's "free" and "rain or shine" attributes. We then emphasized the fact that 90 percent of the Festival's 350-plus exhibits were indoors. We changed our advertising photos to reflect indoor exhibits from previous years and pulled photos that showed blue skies to underscore our message that most of the extraordinary exhibits and presentations took place inside.

Ironically, the *Democrat and Chronicle* for some reason that year had neglected to publish a large series of our digital ads on its homepage earlier in the week. The paper, however, which was a longtime Festival sponsor, apologized and offered us additional free advertising inventory and impressions on the Web and social media in the final 48-hour push leading up to Imagine RIT. This allowed us to amplify our re-tooled message that Imagine RIT would be a great way to spend a rainy spring Saturday.

In the end, the forecast played out as predicted and Festival Day arrived with steady rain and a temperature in the high 40s. As a crowd of about 25,000 to 30,000 "poured" onto campus, Festival Program Chair Heather Cottone exclaimed, "We are now weather tested. And we can handle it!" This was an important test indeed, as beginning in 2018, the Festival moves to a late April date to accommodate RIT's new academic calendar.

IMAGINE THE IMPACT

Andrea Shaver was a high school junior from Frisco, Texas, when she visited RIT in 2012 to participate in a national Young Entrepreneurs Academy competition. When she arrived, Imagine RIT was under way. Shaver, who was seeking a college offering a non-traditional program in art, was immediately taken with RIT's "universe of possibilities" and from that day forward, declared RIT to be the only college she could imagine herself attending.

"I wandered all around campus," she recalled in a 2016 interview with University News, "and saw how students were collaborating with others in different schools and departments. I fell in love with RIT!"

Citing two of the traditional four "Ps" of marketing, Shaver was enthralled with RIT's Product and Placement and enrolled at RIT, eventually becoming both a Festival exhibitor and RIT's Student Government president.

> **Sound Recognition**
>
> The student team of Hz Innovations were confident that they had developed a product that deaf and hard-of-hearing homeowners couldn't possibly live without. A working prototype of their Wavio wireless sound recognition system was on display and in action at 2016's Imagine RIT Festival.
>
> Wavio contains sound-capturing units that are connected to a home Wi-Fi system. When a doorbell rings, a smoke alarm chimes, a water faucet drips, or a dog barks, for example, the unit notifies the homeowner via smartphone, smart watch, tablet, or laptop, and identifies the sound. According to developers, virtually any sound deemed important to the homeowner can be recorded and "memorized" by the system during installation.
>
> At the booth, visitors participated in live demonstrations and offered feedback on the Wavio device. The team collected testimonials from deaf and hard-of-hearing students and homeowners to find out how Wavio could impact their lives.
>
> The team—Greyson Watkins, Chrystal Schlenker, Zach Baltzer, and Nicholas Lamb—won or placed in several local and regional business competitions in 2015-16 and secured a contract to produce 1,000 units. In fact, with product manufacturing a key component to success, the co-founders encouraged students to drop off résumés at their Imagine RIT booth.
>
> Added Baltzer, a fourth-year Microelectronic Engineering student, "It's amazing that a little more than a year ago we were looking for a cool project to focus on. Now we're co-founders of a company."
>
> *Vienna McGrain, University News Services, May 3, 2016*

In 2013, the Imagine RIT Marketing Team won a Gold Medal at the CASE (Council for the Advancement and Support of Education) National Circle of Excellence Awards for its work promoting Imagine RIT. Several of the team's other prestigious awards can also be traced back to successful campaigns in support of Imagine RIT, which has truly become the University's signature branding event. In the process, and as a bonus, the Festival has informed and driven RIT's ingenuity for marketing, public relations, and advertising for other Institute campaigns as well.

Just as President Destler envisioned, the Festival is a catalyst for innovation and entrepreneurship, and a launching pad to jump-start ideas, products, services, and student-led ventures. It also has recruited a new fan in Dr. David Munson, who became RIT's 10th president in July 2017.

"Imagine RIT made an inspiring impression on me," said Munson after attending the 2017 event. "My wife, Nancy, and I observed a sea of students with passion and a strong sense of purpose. RIT students are creators and they know how to put their creativity to use. Imagine RIT shows that creativity and innovation occur in every field and corner of our campus."

imagine *this* 75

David C. Munson, RIT's 10th president, at his first Imagine RIT Festival.
Photo by Brittainy Newman, 2017

In his inaugural speech to students, faculty, and staff in August 2017, Munson articulated his vision to push RIT beyond STEM and career-focused education, to complex problem solving, critical thinking, communication, the performing arts, and the humanities.

The Imagine Marketing and PR team responded to President Munson's cue to double down on efforts to publicize RIT's unique intersection of technology, the arts, and design by continuing to highlight the creative process that is part of RIT's DNA.

We will continue our enterprising strategies and tactics; we will take risks when it comes to trying new methods in media relations, content generation, social media, and advertising. Don't be surprised to see virtual and augmented reality soon finding their way into our approaches. And digital and print (you're holding this book, after all!) strategies will remain relevant. Just as the Festival messaging has said in the past, "What will we think of next?" This drives our team.

Such ideas are part of Imagine RIT's longevity. The Festival continues to expand the boundaries of marketing when it comes to explaining to the public just what makes RIT tick—that collision of "left" and "right" brain sensibilities that both surprise and empower students, faculty, staff, and Festival visitors. As Imagine RIT evolves, so, too, will its marketing/PR stories.

Bob Finnerty *is RIT's Chief Communications Officer. He joined RIT in 2002 and is responsible for enhancing the University's internal and external communications while overseeing University News Services and RIT Sports Information. He has served as the chair of Marketing and Public Relations for Imagine RIT since the inception of the Festival in 2008. Prior to RIT, Finnerty worked as a reporter and editor at the Rochester* Democrat and Chronicle *for 12 years.*

The DNA necklace, intrinsically unique jewelry. Photo by A. Sue Weisler, 2011

PROMOTING THE EXPERIENCE

Fluids are Fun: High-Speed Balloon Drop, American Society of Mechanical Engineers. Photo by Matt Wittmeyer, 2015 top 2, bottom 2010

Max Schulte, 2009

Tom Brenner, 2014

4 MATURATION AND MOMENTUM

HEATHER COTTONE

At the conclusion of Imagine RIT 2015, a modest 6 x 9-inch finger painting, presumably a self-portrait, was left anonymously at a Welcome Tent. The unsigned painting was indeed a mystery, yet it seemed to illustrate perfectly the Festival's intent to inspire creativity. What was it about the Imagine RIT experience that prompted this artist's self-expression? Was the "leave behind" intentional? Indeed it was, as we received a second "mystery" painting the following year, once again unsigned with no explanation of purpose or apparent desire for acknowledgment. As a member of the Festival's Planning Committee, I interpreted it as a kind of thank-you note that, if written, might say, "Your festival is awesome; I had a great time and I love RIT!" But I also wondered: How representative is the young artist's passion of all the Festival's attendees? And if it is so, what can we do to ensure that all visitors' experiences are as uncomplicated and joyfully expressed as the "leave behind" portraits seemed to suggest?

After four successful Festivals, I assumed management responsibilities for Imagine RIT, working closely with Festival Chair Barry Culhane. I had been part of the original Festival planning "team" that created, organized, and propelled the event to adolescence; so, while not as dramatic as being called to the pitcher's mound to preserve a no-hitter, there was a clear expectation of continued excellence that, at times, felt daunting.

President Destler's desire to increase Festival attendance each year until it reached 30,000 was achieved in only three years, much faster than expected. The Festival had also gained an enviable reputation as one of the community's most anticipated, best-attended events. But height and age are only physical measures of maturity; Festival planners now wanted to move the Festival to emotional maturity by enhancing Imagine RIT's experience for exhibitors, volunteers, and attendees.

To keep the Festival's momentum going, we pondered this question: Beyond simple attendance numbers, and moving more deeply into the Festival's working parts, in what other ways could we grow it? Imagine RIT's scope requires nearly a full year of preparation—as one Festival ends, planning for the next begins. So there was plenty of time to imagine, invent, and implement ways to improve an already good thing.

Kate Melton, 2016

STRENGTHENING RELATIONSHIPS

Imagine RIT's intended audiences today remain consistent with those identified by the inaugural Planning Committee in 2007. The importance of the Festival both to and for RIT students, faculty, and staff was critical; were it not for their enthusiasm, participation, and support, the Festival could not thrive as it does. President Destler wanted Imagine RIT to "crack open" the campus so that local and regional community residents might know the wonderful things happening in their backyards. And although Imagine RIT was not created as a recruitment tool, planners would have been naïve to ignore prospective students and their families as a potential audience. Beyond that, organizers sought to attract some of RIT's more than 110,000 alumni spread around the globe, as well as co-op employers from the business community; RIT's cooperative education program is one of the oldest and most respected in the nation.

A. Sue Weisler, 2015

> **Invisible Effects of Emissions**
>
> A typical car emits several metric tons of carbon dioxide and other greenhouse gases each year, but it is difficult for most people to understand fully the impact car emissions have on the environment, because greenhouse gases are invisible to the naked eye. "Gas emissions are a pretty abstract concept for the average person," said Keegan McCoy, an Imaging Science doctoral student. "We are building a system that allows you to see how much gas actually comes out of your tailpipe."
>
> Visitors to 2017's Imagine RIT saw firsthand the effects of car tailpipe emissions using infrared light at the EZ-GAS: Imaging for Car Emissions Characterization exhibit. Twenty-three Ph.D. students from the Chester F. Carlson Center for Imaging Science worked on the project since the fall.
>
> The technology these students are developing is not just for show—they are designing it to be used as a new way to conduct state-mandated vehicle emissions tests. "States are moving toward testing car emissions remotely instead of requiring people to take their car to a local auto shop," said fellow Imaging Science doctoral student Catherine Fromm.
>
> In states such as Virginia, many drivers can complete the emissions inspection requirement for their motor vehicle registration simply by driving through emissions monitoring equipment set up on highway off-ramps. The students used this model as inspiration, but McCoy noted that, "Virginia uses a spectrometer to conduct the test, which is costly technology, but our process is more affordable. We're using basic technology to create a simple solution."
>
> *Luke Auburn, University News Services, April 25, 2017*

Early on, local public, private, and charter schools were made aware of the unique opportunity for K-12 students to attend Imagine RIT; a page on the Festival website was dedicated to "Educators and Mentors" who wanted to know more. Barry Culhane presented an informative talk about Imagine RIT at a meeting of superintendents of Rochester-area schools. And letters later were sent reminding them to "Save the date!" with an offer of flyers and magnets to distribute in their districts.

An additional marketing strategy for the K-12 audience—that continues today—involved emailing teachers to tell them about the year's winning Festival poster. Teachers then could request a poster for their classrooms and were invited to sign up for class field trips via the Festival website. As field trip requests increased each year, Festival planners added a registration form to the site's "Educators" page to collect appropriate contact information, and offer instructions on what educators and students could expect upon arriving at the Festival, including where to park and eat. On Festival Day, these preregistered groups enjoyed speedy, coordinated drop-off and parking services and received Festival programs, promotional items, and easy-to-follow instructions on how to find specific exhibits.

Visualization software recognizes human facial expressions and can assist in future behavioral health therapy.
Photo by A. Sue Weisler, 2013

> **Open Your Eyes and Draw**
>
> Keep your eyes peeled! At the 2012 Imagine RIT Festival, instead of using a mouse to move a computer cursor, you used your eyes and face.
>
> Students and faculty from the Department of Computer Science created systems that captured and used facial expressions and eye gaze to create emotionally expressive artwork. Their exhibit gave visitors a firsthand look at how eye-tracking and facial-expression analysis worked.
>
> "Eye tracking uses infrared illumination and infrared cameras to capture video of your pupil and corneal reflection," said Reynold Bailey, assistant professor of Computer Science. "The video is analyzed in real time to allow you to interact with the screen." Sitting down at the computer system also allows the facial-expression server to recognize lips, eyebrows, and other facial features. A user can then make a specific face that corresponds with a preselected brush stroke. "For example, an angry face will create jagged lines and a happy face will create stars," said Cyprian Tayrien, a Computer Science graduate student. "Then, using your eyes, you can move across the artwork like a normal cursor."
>
> Artists looking to keep evidence of their hands-free drawing simply hit the "save" button with their eyes and had the drawing sent to their e-mail address.
>
> Next door to the drawing station, the Computer Science group also showcased a gaze-based image retrieval system. Similar to the drawings, users searched through a database of images simply by using their eyes. The user could also select different objects or regions in the images by fixating on them. "Looking at a car in one image will cause other images containing cars to be loaded," said Srinivas Sridharan, a Computing and Information Sciences Ph.D. student. "This process can continue until the user is satisfied with their image search."
>
> "Ultimately, systems like this can be used to enable people with disabilities to easily interact with computers using their eyes and facial gestures," Bailey said. Faculty exhibitors included Bailey, Joe Geigel, and Manjeet Rege. Student exhibitors included Sridharan, James Coddington, Junxia Xu, Yuqiong Wang, Bharath Rangamannar, Tayrien, and Stephen Ranger.
>
> *Scott Bureau, University News Services, April 30, 2012*

For several years, a K-12 "Imagine Your Future" game designed for 6- to 15-year olds was distributed at Festival Welcome Centers. The Monopoly-style board game highlighted age-appropriate exhibits located in each of the Festival's "Zones," campus exhibit areas closely aligned with each of RIT's colleges. The game helped children and adult chaperones decide, "Where should we go first?" Children who "played" received attendance stickers or card punches entitling them to giveaway items such as Imagine RIT patches; they were also automatically entered in gift card drawings.

A. Sue Weisler, 2011

Time demands on Festival volunteers necessitated discontinuing the game in 2014; but it was replaced with a survey in which students identified their favorite exhibits and assigned the Festival an overall "grade." In return, respondents received a small promotional item and were automatically entered to win a more substantial RIT Prize Pack. Festival planners used the survey data to learn which exhibits were popular among different age groups, as well as what students said they really wanted to see, such as "more robots" and "more balloons."

While it came as no surprise that children felt free to express their opinions on the survey, what *was* surprising about the K-12 cohort was the age at which these visitors evolved into prospective college students. An affirming testimonial from one set of parents revealed that attending Imagine RIT was a tradition for their family, and that their children—ages 7, 9, and 12—had discovered their career paths while at the Festival. By the Festival's fifth and sixth year, planners also began to notice youngsters who had attended Imagine RIT with their families, returning as interested high schoolers, and later, some began showing up on campus as matriculated RIT students.

As the Festival gained momentum in these first years—both inside and outside the RIT community—organizers also detected a noticeably positive shift in affect about the event among faculty members. Teaching requires detailed planning and adding "one more thing" to an already packed academic schedule—lectures, tests, projects, presentations—isn't always appreciated. But, as Imagine RIT became more successful, faculty members became eager to exhibit their students' classroom projects. This was an important turning point: Instructors appreciated the benefits of having students' work presented to visitors who stopped, asked interesting questions, and prompted students to learn how to respond "on the fly." Likewise, student exhibitors expressed appreciation for what they called "real world" interactions and feedback.

AstroDance, multimedia performance using dance to tell the story of black holes and the gravitational pull of particles in the universe, a joint project among the College of Science, the B. Thomas Golisano College of Computing and Information Sciences, and NTID's Performing Arts program. Photos by Mark Benjamin, 2013

Festival organizers also began receiving an increasing number of requests to hold events as add-ons or in-conjunction-with Imagine RIT's Saturday slot. One memorable request was an RIT student club that wanted to hold its anniversary celebration—which would include both students and alumni—the same weekend as Imagine RIT. While such requests were flattering, people and groups using Imagine RIT for what we began to call "satellite" events also stretched the resources of Festival planners significantly. Organizers did their best to accommodate such requests, but some individuals and groups had to hold and manage their events on their own.

Because RIT's long history has yielded a substantial cadre of loyal alumni, RIT's Office of Development and Alumni Relations saw Imagine RIT as a special opportunity to connect with this important cohort, especially those living in the Northeast region of the United States. In 2012, an "Alumni Oasis" was set up inside Booth Hall's University Gallery; it was later relocated to the lobby of the newly constructed Gene Polisseni Center. Alumni who pre-registered to attend Imagine RIT received a gift, free food and beverages, and a list of family-friendly activities to enjoy, such as a photo booth and face painting.

The area proved popular instantly, with nearly 1,000 alumni and their guests visiting during each Festival. Alumni Relations staff energetically reached out to engage the alumni visitors, many of whom had not been on campus in years, or since the previous year's Festival.

EXHIBITORS AND VOLUNTEERS

In 2012, Festival planners partnered with RIT's Information and Technology Services Department to develop a more comprehensive and user-friendly online exhibit proposal system. After four years, planners had learned the "right" questions to ask potential exhibitors and with a deeper level of detail. One example was creating a more efficient way for exhibitors to upload or select photographs to represent their exhibit descriptions on the Festival website. Another was greater awareness of health and safety issues, such as asking whether potential exhibits would incorporate things such as propane, lasers, or live organisms. Festival planners also learned a great deal about fire safety regulations for events taking place inside buildings —for instance, no helium balloons are allowed in the Gordon Field House, since a wayward balloon would trip the fire alarm and sprinkler system. Nor could exhibits be set up in a building's lobby—too many visitors would block the flow of foot traffic and impede a safe exit in the event of fire.

A. Sue Weisler, 2015

Organizers used the Imagine RIT online portal to email exhibitors about checklist items such as proposal deadlines, exhibit placement, internet networking needs, and how to pick up Imagine RIT T-shirts and meal vouchers—the former for publicity and identification purposes and the latter for the always-present hunger pangs of college students.

As the size of the Festival grew, so did the energetic and indispensable cadre of volunteers. RIT staff—everyone from groundskeepers to the custodial crew, from office managers to staff assistants—are the reasons why Imagine RIT works. Altogether, they number in the hundreds and their involvement remains crucial to the Festival's success. Not only do RIT staff have their fingers on the pulse of the Institute, they have the pragmatic know-how that allows strangers to the campus—such as most Imagine RIT attendees—to get from one place to the next, including back

Matt Wittmeyer, 2011

to their cars. With hundreds of exhibitors spread out across a growing campus footprint as the Festival evolved over the years, Imagine RIT's continued success was and is dependent on the more than 500 volunteers who enthusiastically greet, assist, and direct visitors.

Over the years, many of this army of way-finders and problem-solvers were RIT students as well as staff. As the critical "first touchpoint" for Imagine RIT's visitors, they populated the Festival's Welcome Centers and Information Booths. Others greeted visitors, offered directions, or drove golf carts to assist those who needed a lift. Better than any sign could ever hope to be, the volunteers humanized the Festival with their knowledge about the campus, its history, and, as it developed, the Festival itself.

Brett Carlsen, 2014

Blü Bloat, 2015

Mark Benjamin, 2011

NTID Performing Arts, theatrical productions presented in spoken English and American Sign Language. Photos by Mark Benjamin, 2009 and 2011

> **Domino Daze**
>
> The RIT Tiger logo came to life in dominoes at the 2015 Imagine RIT at the "Tigers Earning Their Spots" exhibit created by PiRIT, the RIT Association of Student Mathematicians and Statisticians. Visitors to the PiRIT booth in Gosnell Hall were likely unaware of the mathematical concepts lurking behind the eight-foot by eight-foot portrait composed of 81 sets of dominoes, or 2,268 individual tiles.
>
> PiRIT vice president Tamalika Mukherjee, a fourth-year BS/MS student in Applied and Computational Mathematics, encouraged visitors to come a little closer to learn how the students arranged the dominoes. "We wanted people to see how advanced mathematical topics like graph theory and linear algebra could be applied to create something tangible and artistic," Mukherjee said. "We explained the math behind the algorithm generating the domino portraits in more detail at our exhibit."
>
> The PiRIT team depicted the large-scale logo in dominoes using an integer linear programming model proposed by Robert Bosch, professor of mathematics at Oberlin College. Ten members of PiRIT followed the algorithmic blueprint and laid down domino after domino. "We were really excited about creating this exhibit," said Christine van Oostendorp, president of PiRIT. "Sometimes it can be difficult to convey mathematical concepts. Imagine RIT offered an opportunity to show that all the computation and numbers can result in an eye-catching product. Last year, our Sierpinski Triangle was definitely a focal point for the College of Science, and I know many are looking to see what we do next."
>
> Visitors saw for themselves how math and art intersect by building smaller domino pictures of tiger paw prints using the same concepts applied to the massive RIT logo. "We wanted to scale the size down so that people who visited our exhibit could build their own domino portrait and take it home with them," Mukherjee said.
>
> In addition to Mukherjee and van Oostendorp, exhibitors included students Ellen Baillie, Kyle Cutler, Ben Evans, Renee Meinhold, Erin Neidhart, Eric Peterson, Mike Spink, and Evan Witz; and, School of Mathematical Sciences Professors Michael Cromer, Matthew Hoffman, and Paul Wenger.
>
> *Susan Gawlowicz, University News Services, April 29, 2015*

SPONSORS

Imagine RIT's first major sponsor, in 2008, was Arunas Chesonis, then CEO of the Fortune 1000 telecommunications company, PAETEC, and a member of the RIT Board of Trustees. After Barry Culhane presented plans for Imagine RIT to Trustees and described the financial support needed, Chesonis approached him and, with a gentleman's handshake, became the first "premier" level sponsor for Imagine RIT.

Following in PAETEC's footsteps, other major Festival partners included the Rochester *Democrat and Chronicle*, Kodak, Paychex, Toyota, Xerox, and Rochester Regional Health System. Perhaps the longest standing Festival partner was Time Warner Cable Company, which from 2010 through 2016 was a premier sponsor. The "marriage" was fortuitous, as Time Warner had just launched a philanthropic initiative called Connect A Million Minds, to address America's declining proficiencies

in science, technology, engineering, and mathematics—and Imagine RIT was a perfect fit for that message. Our champion within Time Warner Cable was Terence Rafferty, a fellow volunteer and friend of Barry at the Al Sigl Community of Agencies.

FESTIVAL CONTESTS

In addition to exhibiting during the inaugural Festival in 2008, RIT students were also invited to design an official Imagine RIT Poster. The poster contest, with its prize of $500 for the winner and $250 for the runner up, was intended to market the Festival among RIT's population. The winning design was printed as a 14 x 20-inch poster distributed free at a Festival Poster Tent set up outside The Wallace Center, enhancing and extending Imagine RIT's post-event exposure. Some people have complete collections of the posters in their homes and offices.

The first year, organizers received a dozen or so poster submissions. Today, as many as 75 might arrive. Entries are posted in an online gallery on the Imagine RIT website, where fans have one week to vote for their favorites. Voting is publicized through email and social media channels. The contest fuels friendly "water cooler debates" in buildings around campus before the Festival; and not surprisingly, also spawns competition among students. RIT students are justifiably known for their technological prowess, and finding a "workaround" for an online voting system is low-hanging fruit. Fortunately, patterns of ballot stuffing have been easily discernable to equally savvy Festival planners.

Over the years, organizers sponsored other contests to generate enthusiasm within the RIT community—contests that were designed for one-time use. To promote the inaugural Festival, for example, students created 30- to 60- second YouTube videos; another year, RIT's a cappella groups wrote and performed an original song about the Festival. A social media video contest in 2011 was won by a team of students who created a trio of videos—each featuring 3D digital graphics—that included the bubbles from Imagine RIT's logo and a bouncing brick.

imagine *this* 103

DR. DESTLER'S CHALLENGE

From 2009 through 2017, Dr. Destler's Challenge was among the most anticipated events preceding the official opening of Imagine RIT. Early challenges involved different "green" e-vehicles racing on short courses set up either in Simone Plaza, along Andrews Drive, or in a campus parking lot. Races usually began at 8 a.m. on Festival Day; winners were announced during the Festival's 10 a.m. opening ceremony, which typically consisted of brief remarks by President Destler, RIT's Board Chair, and the CEO or President of the Festival's premier sponsor. Many Destler Challenges required students—with input and oversight from faculty and staff— to modify existing vehicles, such as Power Wheels toys, go-karts, utility carts, or bicycles. A few, however, involved students building vehicles from the ground up. Competitor scores on several criteria determined the winner: Who completed the

Time Warner Cable, a premiere sponsor for Imagine RIT from 2010 to 2016, launched its philanthropic initiative "Connect A Million Minds" to address declining proficiencies in STEM education.
Photo by A. Sue Weisler, 2014

course fastest and used the least amount of energy? Which vehicle traveled the farthest? Since the scoring criteria changed depending on the event, the competition was always fresh and entertaining.

In 2015, Dr. Destler's Challenge was an Unmanned Aerial Vehicle—aka drone—competition held in the newly constructed Gene Polisseni Center. The next year, the challenge focused on accessible and inclusive technology research—a Mobile Pediatric Stander, an electric wheelchair in which the user can transition to an upright position, and an Intelligent Mobility Smartcane for those with visual impairments, were among the creations. Winners of the 2017 Challenge, which focused on Health Innovations, included a Tremor Mitigation Device and an athletic shirt that sends quantitative workout data to the wearer's phone.

Bill Destler with an antique banjo from his collection.
Photo by A. Sue Weisler, 2013

A. Sue Weisler, 2015

President Destler enjoyed and often participated in the competitions. One year he used his own custom-built electric bicycle, which he rode to campus on most good weather days. Dr. Destler also took part in the UAV competition, although his drone had a disastrous crash. Winners of Dr. Destler's Challenge could choose either a $1,000 cash prize or one of the president's antique banjos from his collection of more than 150 world-class instruments; most chose the banjo.

MASSAGING THE MESSAGE

In 2012, the Festival underwent a marketing "facelift" to keep it modern and relevant. A new television commercial leveraged images from past Festivals and a website redesign freshened the Festival's public face. The site further refined its intelligent Plan Your Day tool, with an Amazon-like section that used a visitor's expressed areas of exhibit interest to recommend other similar Festival exhibits. In other words, people who said they were interested in the S'more Manufacturing Exhibit were offered suggestions for other engineering or food-related exhibits. With more than 400 exhibits to choose from, this strategy helped prevent visitors from feeling overwhelmed and allowed them to choose to either wander the campus and enjoy a chance encounter with an exhibit, or "steer" their experience through a pre-selected itinerary. Today, Imagine RIT continues to deliver on everything its name suggests: Innovation, creativity, and fun. Festival organizers enjoy their running conversations with attendees, exhibitors, and volunteers during the Festival; and communicate with their many audiences afterward through debriefings and post-Festival online surveys.

After 10 successful years, perhaps the event's most important takeaway is this: Imagine RIT no longer is marketed as an *event*. It is now part of the University's *brand*—promoting Imagine RIT means promoting *RIT*.

After Imagine RIT 2017, another small painting was left at the Festival Welcome Center and eventually made its way to my office. I posted the image on Imagine RIT's Facebook page, asking: "Does anyone recognize the art or artist? Where is he/she from?" And perhaps most importantly, "Why is this artist leaving behind his/her artwork every year?" Within a few minutes, I heard from a teacher in a nearby rural community who confirmed the artist's name and shared some general information about him. I do not yet know how old he is, but I like to think that maybe I'll meet him at an upcoming Imagine RIT Festival—and that he will introduce himself to me as an RIT student.

Heather Cottone *is Manager of Special Projects for RIT's Office of the President and Chairperson of Imagine RIT's Program and Logistics Committees.*

Kate Melton, 2016

Matt Wittmeyer, 2015

imagine *this* 109

Oobleck Trough exhibit, Kate Gleason College of Engineering. Kids walk on the liquid layer of corn starch and water mixture that turns solid upon contact with an external force. If they're fast enough, they're walking on water but, if they stop, they sink.
Photo by Matt Wittmeyer, 2012

110 MATURATION AND MOMENTUM

A milk jug igloo, Re-Packaging for Tomorrow exhibit
Photo by Mike Bradley, 2015

A. Sue Weisler, 2011

Kedar Mahajan
Class of 2016
GO GREEN!!
Kyla Chiaren
Katrina Harbeson
Emery Martz Will Go to college here
Karan Schroeder
IPKP 98
Imani Was Here
MARA Golden
Olivia was here
Cydney Owens
-Melissa AKA Charmander
Erica
Amari

Wrist-driven Osprey Hand, e-NABLE the Future exhibit.
Photo by A. Sue Weisler, 2016

imagine *this* 113

A. Sue Weisler, 2011

A. Sue Weisler, 2011

5 AT THE INTERSECTION OF LEFT AND RIGHT BRAIN

JEFF SPEVAK

The students and their professors work in the clean, well-lit environment of a modern college campus. If it is a beautiful afternoon, they will almost certainly pick up their paper coffee cups and move outside, perhaps sitting in the shadow of *Sentinel*, a public sculpture whose meaning is as inscrutable as the Sphinx.

There, these Rochester Institute of Technology students can power up their devices and imagine: If today's roads could be navigated by driverless cars, what other uses might exist for that technology? Is it possible to install a microphone and that same driverless-car software on a standard wheelchair, enabling its user to issue voice commands to navigate the chair? Can electrodes attached to the head convert a neurologically challenged person's facial movements into signals that likewise control a wheelchair? Can a computer tablet be reconfigured to enhance communication among deaf, hard-of-hearing, and hearing people? Is one small answer to ecological sustainability the "backyard biodigester," a system that consumes the remnants of your chicken dinner, human waste, and diapers and converts it into fuel? How about a 3D-printer that can create a prosthetic leg for a dog? A computer-guided glove that controls involuntary tremors in the forearm?

In its first decade, Imagine RIT: Innovation + Creativity Festival has explored all of these ideas. It is a festival designed not only to emphasize practical thinking on the part of the school's students, but perhaps to create something that will enhance quality of life. Incremental advances, moving humanity forward.

"Festivals and events always have been able to present new ideas," says Erica Fee, producer of the KeyBank Rochester Fringe Festival. Congregations of people, Fee contends, are "transformative in nature," making Imagine RIT "the modern day, tech version of the groundbreaking Great Exhibition of 1851."

Google it. Considered the first World's Fair, sometimes referred to as the Crystal Palace Exhibition, this was a gathering in London's Hyde Park of the world's elite people and ideas and hundreds of thousands who wanted to see what the future might hold. Charles Darwin was there, and Charlotte Brontë and Charles Dickens.

PlayMobile: Motorized Pediatric Stander, Kate Gleason College of Engineering.
Photo by A. Sue Weisler, 2015

They witnessed miracles such as Frederick Bakewell's "image telegraph," which after another 125 years of tinkering evolved into the fax machine. Even the look of the Great Exhibition was a vision of things to come: Its main exhibition hall, The Crystal Palace, abandoned traditional stone and brick construction in favor of a cast-iron framework embracing a new building material, large sheets of plate glass. The Crystal Palace is now gone—its death was inevitable—but illustrations of it from those days look like pages from a promoter's broadside announcing the arrival of the Industrial Revolution.

Awe-inspiring architecture often falls victim to time and advances in engineering. But just as dinosaurs gave way to mammals, more efficient designs emerge and take hold in even larger ways. When groups of people gather to witness the new idea's superiority, even a seemingly modest innovation can move as swiftly as Laura

A. Sue Weisler, 2016

Intelligent Mobility Cane with Navigation, an assistive device for blind and deaf-blind people, Golisano College of Computing and Information Sciences.
Photo by A. Sue Weisler, 2015

Piloting a quadcopter during a competition.
Photo by Mike Bradley, 2015

> **Hot Wheelz Takes Flag**
>
> The e-dragster team captain was aptly named Ferrari. The driver of the speedy electric vehicle raced cars and motorcycles growing up.
>
> Both were part of the nine-member, all-female Hot Wheelz team that went 100 meters in just under six seconds to win the E-Dragster Race at the 2012 Imagine RIT: Innovation + Creativity Festival. More than a dozen teams competed in the electric-powered vehicle drag race against RIT President Bill Destler.
>
> Students from the Women in Engineering program formed Hot Wheelz and built a lightweight go-kart chassis fitted with six AGM-lead batteries to power the 72-volt, 10-horsepower motor that gave them the edge in the competition.
>
> Driver Maura Chmielowiec, a first-year Mechanical Engineering student, reached a top speed of 58 miles per hour. "Doing this," she said, "I got to see that my mechanical engineering knowledge can be paired with electrical to create something really great—and absolutely fast." Natalie Ferrari, team captain, agreed. "We kept tweaking and tweaking the car the last few weeks. It really paid off," said the Mechanical Engineering major.
>
> The winning team chose an antique banjo from Destler's personal collection instead of $1,000 for the grand prize, saying the banjo was something they could keep and display to remember the victory.
>
> *Michelle Cometa, University News Services, August 6, 2012*

Ingalls running from a prairie fire. Fee points to two inventions in particular from World's Fairs that took different routes to the public. What was known as the "clasp locker" was introduced in 1893 in Chicago. Like the image telegraph, it didn't quite catch on until years later, after it had evolved into the zipper. Yet another creation had been around for a while before the public developed a taste for it at the 1904 Fair in St. Louis: the ice cream cone. Serious or frivolous, the zipper and the ice cream cone are inventions that are now so common that one hardly gives them a second thought after pulling on one's pants or taking that first lick.

Although Rochester doesn't have a lock on failed public events—barbecue competitions come and go everywhere—Rochester's history with festivals has been spotty. The 2015 "A Street Light Festival," with its promises of glowing dancers and 10-foot cubes with interior illumination interacting with patrons like a house cat, came off like the Charlie Brown Christmas tree. Yet there are solid successes to be found, most notably in the internationally acclaimed Lilac Festival and the world-class Xerox Rochester International Jazz Festival. The Lilac Fest was born in 1908, evolving over time from a shrub-sniffing daytime stroll through Highland Park to evenings of thousands of people spaghetti dancing in front of the Pittsburgh world-groove band Rusted Root. And the Jazz Festival, launched in 2002, grew into one of the world's top events of its kind, drawing 200,000 people each year for acts ranging from Jeff Beck to Diana Krall to Woody Allen's New Orleans Jazz Band to Faroe Islanders singing about crows. After its first decade, it was clear that Imagine RIT had joined this elite group.

Brett Carlsen, 2014

Was there room for it? Rochester's summer, which seems to have expanded as climate change settles in, has always been filled with community happenings, with more spilling into the remaining three seasons, elbowing each other for space on the calendar. Celebrations of ethnic cultures, music, food, and beer. Film festivals focusing on women, gay issues, the environment, and movies presented on nitrate stock. Three of the largest and most-established festivals are geared almost exclusively to attracting vast pilgrimages of crafts seekers. July's Corn Hill Arts Festival sprouted in 1968 in the same neighborhood that once had been home to Shop One, which rejected the '50s mania for mass-produced art in favor of hand-produced craftsmanship. As an outlet for the work of acclaimed RIT ceramics professors such as Frans Wildenhain and Hobart Cowles, Shop One was one of the country's most unusual arts stores, and an excellent example of an area's bastion of higher education integrating with its supporting city. Completing the triptych are August's mile-long Park Avenue Summer Art Festival, which joined the lineup in 1976; and September's Clothesline Festival, established in 1956 and a major fundraiser for the Memorial Art Gallery. All commercial in nature—unlike Imagine RIT—their transformative potential is limited mainly to shoppers clutching their newfound treasures.

Yet art as a major driver of festivals is not limited to pursuit of the perfect bird feeder or garden ornament. The performing arts figure into this formula as well. And here, we begin to see a fusion of technology and art. One of the best-known examples cited by Fee is South by Southwest (SXSW) in Austin, Texas. Born as a music festival, after a few years it not only was the most important industry conference in the nation, but also had begun sprouting more components—a film festival and then an interactive convention built on the promise of the internet. SXSW evolved into a perfect ménage à trois of media, especially with respect to social media, "where you saw all of these ideas that had been simmering in labs for years," Fee says.

Mark Benjamin, 2011

Her Rochester Fringe Festival was modeled after similar events around the world, most notably the Edinburgh Festival Fringe. As is the case with most festivals of the fringe nature, the Rochester Fringe sought to promote performances and art that are both mainstream and daring: From improvisational comedy groups riffing on the apocalypse and historical examinations of the Holocaust, to drag queens exploring the magic of Liza Minnelli. By 2018, it had grown to 11 days in September, a scheduling calculation intended to draw on the approximately 33,500 students from the area's four-year colleges. Early on, RIT students and professors were a particularly strong presence. Perhaps this interaction with a city festival is a spillover

Brett Carlsen, 2014

from Imagine RIT. Future engineers searching for the muse. "Imagine RIT is a very exciting thing with a lot of energy," Fee says. Imagine RIT was where she first saw "wearable tech," light-emitting diodes worked into the design of clothing, which first appeared at Rochester Fringe in its early years. "LED fashion," Fee says. "At the time, it was really groundbreaking."

Matt Wittmeyer, 2014

Shake, Rattle, and Roll

How many times has a package arrived through the mail looking like it was on the field during a World Cup match? Did the product inside survive?

Visitors to the 2015 Imagine RIT Festival learned how packaging products are tested to endure challenges along the route from store to home at the "Packaging—Will It Survive?" exhibit. Faculty and students in RIT's Packaging Science Department demonstrated how packages are tested in simulated distribution and design situations in its Dynamics Lab, located in Louise Slaughter Hall. Drops, shocks, and vibration demos took place throughout the day.

The Dynamics Lab is certified as an International Safe Transit Association Test Lab with equipment to test for shock, vibration, and compression and how these parameters might affect packages and products. In addition to being a teaching lab, the facility is used by industry for testing.

Once packages are delivered, though, might they have a second life? Emphasizing sustainable packaging options in its program, the Packaging Science Department displayed alternatives as part of its "Re-Packaging for Tomorrow" initiative. A poster exhibit of Packaging for Tomorrow and Beyond was a virtual trip to the year 2050 and showed how packaging could interface with the smart living spaces of the future. The Re-Packaging for Tomorrow Project display and activities included arts and crafts using recycled materials and an igloo made of milk jugs.

"Re-Packaging for Tomorrow advocated sustainability and up-cycling of wasted materials," said Yuwei Qaio, a second-year Packaging Science student. "Participants learned better ideas of how wasted packages can be converted to craft projects or home decorations," she said, adding that volunteers from the Packaging program collected the materials and participants at the Imagine RIT Festival made their own crafts from water or juice bottles, glass bottles, or cereal boxes. The same student volunteers collected more than 500 milk jugs to construct a sustainable igloo.

Michelle Cometa, University News Services, April 23, 2015

And this is the intersection at which we find Imagine RIT. In 2007, the new president, Bill Destler, brought to the school a background of research into high-power microwave sources and advanced accelerator technologies. He also had a curious interest in stringed musical instruments, with a collection of more than 160 vintage banjos. He even released a folk album, *September Sky*, in 1973. One of his first acts at the school was to create the Imagine RIT Festival. Destler's successor, David Munson, arrived in 2017 with a résumé that also leaned right brain. "My inclination is to work at the intersection of technology and the arts," Munson said. While dean at the University of Michigan, he linked electrical engineering and computer sciences with the arts schools in a concept called ArtsEngine. Creativity and practicality driving each other. RIT's legacy of combining technology and the arts appears safe.

StrongArm was originally developed to help support posture for heavy lifting with less risk for injury.
Photo by A. Sue Weisler, 2011

Silently watching over this collaboration is Albert Paley's *Sentinel*. Paley is one of the world's foremost metal sculptors, whether the discussion is about a pair of $14,000 candlesticks or massive pieces that transform public spaces. A Philadelphia native, he got his start at RIT in 1969, teaching goldsmithing, before moving on to focus on his sculptures. At his studio just a few miles from the school, new technologies coexist with the old. Upstairs is sort of a working Paley museum, including pencil sketches, some framed, hand-drawn on extra-large sheets of white paper. And cardboard mockups of what is to be, maquettes, complete with tiny human figures to suggest the scale. There are computers and software technologies as well, that RIT students would recognize and quickly feel comfortable operating. These are not for Paley, but his 12 full-time employees. "I've never touched a computer in my life," he says. Downstairs is a 32,000-square foot space that looks less like an art studio than a warehouse where bridges are being assembled. A world of forges, plasma

A. Sue Weisler, 2010

cutters, hydraulic presses, and rollers that bend steel into art. And work benches, anvils, and dozens of hammers, many 100 years old. "The hammer I use was made in 1905," Paley says. This may be the 21st century, but the art of blacksmithing survives.

Paley's studio is a vivid example of the cross-pollination of technology and art. Imagine RIT, as might have been said by Aldous Huxley—in a concept later co-opted by The Doors' Jim Morrison—opens the doors of perception, so that the learning experience embraces equal range. Yet the classroom as laboratory is not enough. Fee offers a quote from playwright and director David Mamet's book *True and False: Heresy and Common Sense for the Actor*. "The audience will teach you how to act and the audience will teach you how to write and to direct. The classroom will teach you how to obey, and obedience in the theatre will get you nowhere. It's a soothing falsity."

Kyle Hofsass, 2015

Stepping through those same doors into Imagine RIT, students and visitors are now out of the comfort zone of the classroom and home. "Education by nature is one of transformation," Paley says. "In the process of education, they change, see things they never saw before. That's the whole process of art, by and large. The whole idea is not to do what you know. But experience what you don't know."

Imagine those student inventors and student conjurers, their plans coalescing over coffee and laptops, as they glance up at *Sentinel*, in whose shadow they sit. The towering sculpture looks like many things, depending on the angle from which it is viewed, how the sun strikes its surfaces, and the shadows that are cast. A sheaf of steel shapes, chaotic yet working together.

Technology and need—and sometimes war and greed—are the prime drivers of invention. But art inspires, prompts, and fuels new ideas as well. In this respect, Imagine RIT became a perfect fit for a school geared for unleashing humans who work both sides of the brain. Science and mathematics on the right side, creativity and the arts on the left. If such a differentiation really exists. Cognitive neuroscientists disagree, but that's a debate for another book. And as a metaphor, right-brain left-brain works just fine for Imagine RIT.

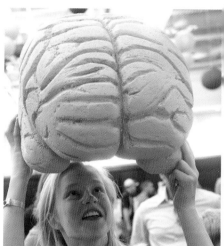

Jeff Spevak, longtime entertainment writer in Rochester, N.Y., has interviewed Johnny Cash, Patti Smith, Tony Bennett, James Brown, Noam Chomsky, and Mickey Mantle. But his biggest regret is turning down an opportunity to hang out with Willie Nelson on his tour bus.

imagine *this* 133

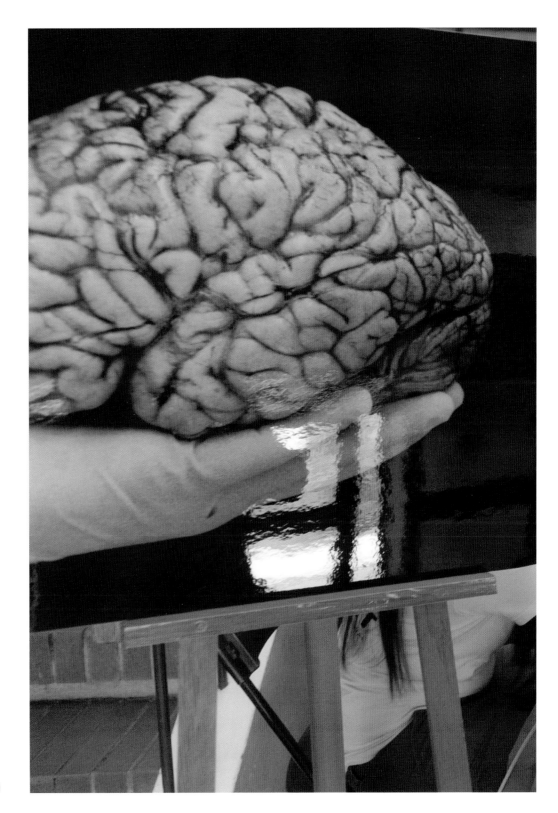

A. Sue Weisler, 2013

A. Sue Weisler, 2010

A. Sue Weisler, 2015

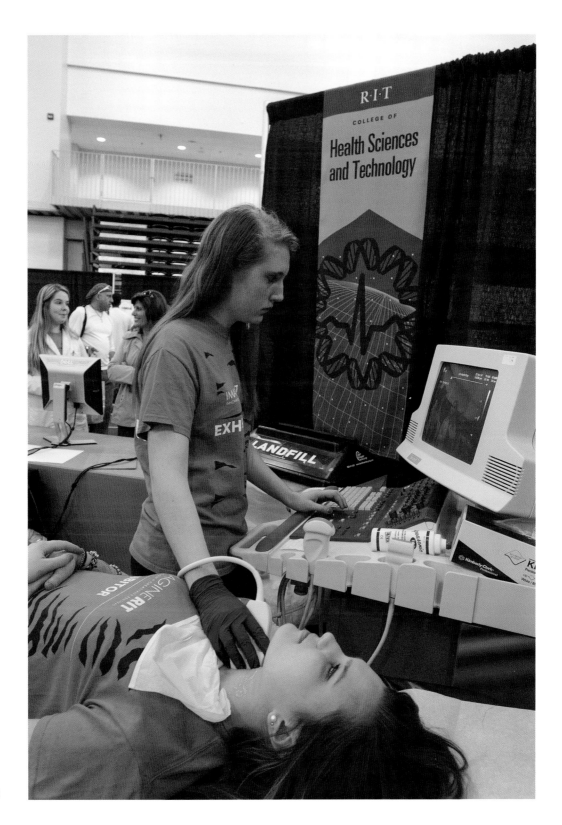

A. Sue Weisler, 2014

AT THE INTERSECTION OF LEFT AND RIGHT BRAIN

Matt Wittmeyer, 2015

A. Sue Weisler, 2015

AT THE INTERSECTION OF LEFT AND RIGHT BRAIN

Mike Bradley, 2017

A. Sue Weisler, 2012

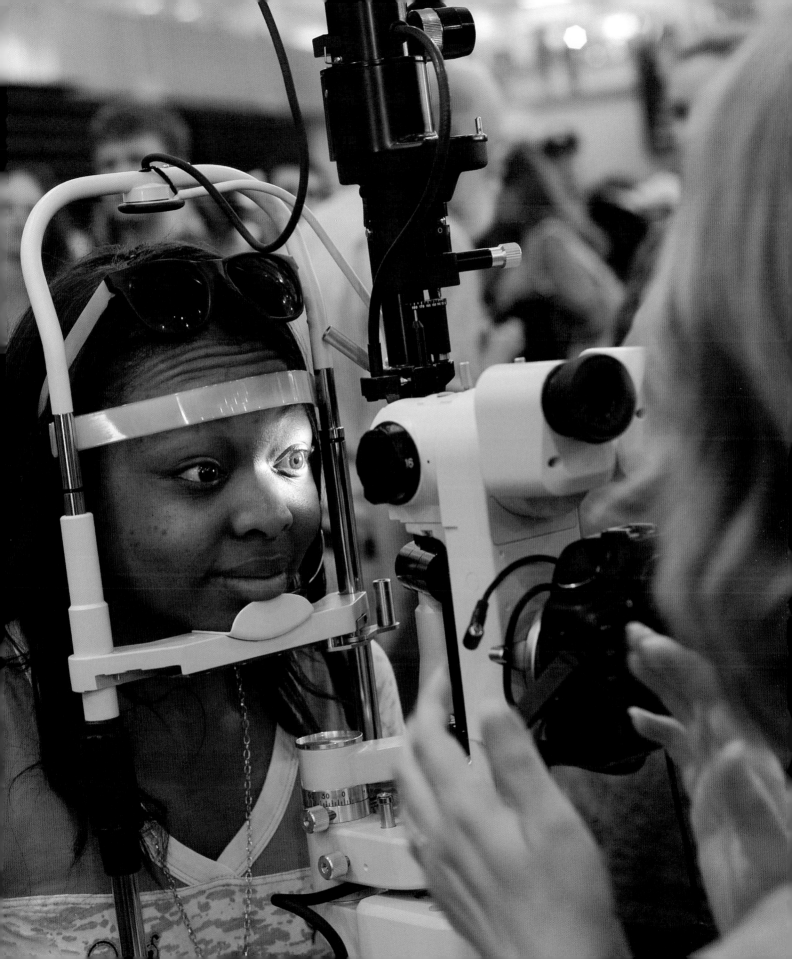

AT THE INTERSECTION OF LEFT AND RIGHT BRAIN

The ThermApparel team won an Al Sigl prize as part of Dr. Destler's Access and Inclusion Technology Challenge.
Photo by A. Sue Weisler, 2016

A. Sue Weisler, 2014

A. Sue Weisler, 2013

6 A GALVANIZING FORCE

BILL DESTLER

Imagine RIT has educated and delighted thousands of visitors from around the country for 10 years, growing in stature to become the University's signature event. Originally, it was conceived as an event significant in terms of external PR and reputation-building value. What no one realized was how important Imagine RIT became internally at RIT.

It is amazing how little convincing was needed to get the idea off the ground. Barry Culhane and Andrew Quagliata deserve mountains of credit for the way in which they doggedly pursued the details—where, indeed, the devil often resided—while generating tremendous enthusiasm and pride among students, faculty, and staff members. Barry knew who to ask and what to ask for to get things running; I held the enviable position of being "the new guy" who watched with admiration and awe as the Festival brought the campus community together in ways no one anticipated.

A short while before that first Festival began, Andrew and I looked wistfully toward RIT's Jefferson Road entrance, hoping to see at least a few cars turning in. They did, of course, but there were more than a few anxious moments. Those moments were replaced with sheer delight, however, as we walked around campus and saw firsthand the innovative work on display, watched the delighted responses of the crowds, and appreciated the cross-college collaborations the Festival had spawned.

So while we now chuckle at some of those "beginner's mistakes"—from over-taxed electrical outlets to underprepared food vendors—the post-Festival glow that enveloped the campus far outshone the occasional glitch or gaffe.

After 10 years, I remain mystified by some of the projects. One that comes to mind involved first-year Imaging Science students who produced a video showing a light going on in a room and the gradual illumination of the room as the light traveled from the bulb to the walls. I still have no idea how they did that. And perhaps the Festival's craziest idea—which was mine—was to hold an electric vehicle "drag race" in a campus parking lot. When one of the student participants excitedly told me that he believed his vehicle could reach 110 mph over the planned quarter-mile course, we realized the safety issues associated with that possibility and immediately shortened the course. Still, the fastest vehicle reached a top speed of 70 mph—matching how quickly I exhaled once the competition was safely over.

A. Sue Weisler, 2010

A. Sue Weisler, 2009

Imagine RIT has been a galvanizing force for innovative and creative work at RIT. After a year or two, students, faculty, and staff started to plan their participation in the event at the beginning of the academic year, and many classes changed their due dates for significant projects so that students could exhibit the results of their work at the Festival.

While much has been written touting Imagine RIT as my personal "legacy," nothing could be further from the truth. The Festival's success belongs, instead, to the thousands of genuinely good and creative people from RIT who turned an idea into an overwhelmingly successful reality. Our goal always was to celebrate students' creativity in a grand, inclusive event where "everyone could find a place under the tent." I believe we succeeded.

In some ways, the process of creating a festival of Imagine RIT's magnitude is similar to that of building a musical instrument—an analogy that won't surprise those who know my affinity for banjo playing and collecting. Both require vision, patience, expertise, and some kind of "glue"—real or metaphoric—to hold the idea together. Here's to all who have worked to create the Imagine RIT Festival—a beautiful instrument of another kind, built to last.

Bill Destler, *Rochester Institute of Technology's ninth president, 2007 to 2017, was first to imagine Imagine RIT.*

Bill Destler listening to the RIT Pep Band.
Photo by A. Sue Weisler, 2010

A. Sue Weisler, 2010

imagine *this*

A. Sue Weisler, 2014

A. Sue Weisler, 2008

imagine *this* 153

Matt Wittmeyer, 2011

Mike Bradley, 2016

A. Sue Weisler, 2013

COLOPHON

DESIGN
   Lisa J. Mauro

PRODUCTION SUPERVISION
   Marnie Soom

PRINTING AND BINDING
   Thomson-Shore, Dexter, Michigan

PAPER
   Sappi Flo Matte, 100# Text

TYPEFACES
   Glypha by Adrian Frutiger (D. Stempel AG, 1977)
   Univers by Adrian Frutiger (Deberny & Peignot, 1957)